Efrain Cordero's

EUROPA

(The City of Fury)

outskirtspress

DENVER, COLORADO

EUROPA
The City of Fury
All Rights Reserved.
Copyright © 2012 Efrain Cordero
v2.0

Outskirts Press, Inc.
http://www.outskirtspress.com

ISBN: 978-1-4327-7384-7

Outskirts Press and the "OP" logo are trademarks belonging to Outskirts Press, Inc.

PRINTED IN THE UNITED STATES OF AMERICA

Introduction

The best place to find life in our solar system is Europa, one of Jupiter's four large moons. It has a nearly planet-wide saltwater ocean buried beneath several miles of ice and kept warm by gravitational tidal flow. Europa may be the only place in our solar system besides Earth that contains a great deal of water, researchers say: Gravity and magnetic data collected by NASA's Galileo orbiter have provided increasing evidence that an ocean exists underneath Europa's uniform (10-100 km thick) coat of ice. The possible ocean on Europa may contain more liquid water than all the oceans on Earth combined. Researchers hope to discover whether Europa is made up entirely of ice or if it contains an ocean underneath (in which if there's water there may be life…).

In the year of 1977, there was a strange event that caused a commotion for the people in Central America, on a dark night of summer between 7 and 7:30 p.m. The sky lightened during the space of three to four seconds; it looked like it was already daylight, but nobody knew what happened exactly that day. There were some witnesses who described a spacecraft coming from the sky. It landed close to a deserted forest, and in a few minutes it was covered by soldiers from the US Army who found three aliens inside of that ship (two men and one woman). Their government denied it of course, but they were escorted into a facility right away. They kept the aliens as prisoners, experimented on their DNA, and kept the event the whole time a secret until now. The main reason these aliens came to our planet was to tell us how to prevent a big invasion from Europa; of course our government didn't believe it. In their spacecraft they were carrying a mineral called Super Eurotanius, which was a super mineral never found in our solar system but only in Europa. But some people of that facility saw something extraordinary in those aliens and helped them to escape from that place. They were taken to different countries just to get rid of their hunters, and they were placed in El Salvador and Alaska. They were not found until the unexpected happened, but it was too late. Earth had been attacked by powerful bombs. They were targeted by the Zurdos, a race of hunters living in Europa. One of the great cities in Europe had been wiped out, and the same in America, so it was time to locate the first three emissaries from Europa. People were living in chaos, waiting for the final days.

On Earth they needed them so badly that our planet was at risk of being wiped out. But after many years, a lot of things had changed. Ingui, Une, and Currumiche (the aliens) were not in position to fight; they were not the same as when they were younger but they had something else—their sons, Alexandrus, Dulcinea, and Amadeus. They were the only hope that Earth had.

Back in El Salvador

Lt. Dago from the US Army was in ESA trying to find Alexandrus, the alien. The planet Earth was in chaos, and we were very surprised about the existence of other human forms in our galaxy. We were warned but we did not believe it. He looked for so long everywhere without success until one night he went out to a night club and that's when it all started! He went to the bartender.

Lt. Dago: Disculpe, usted ha visto a el.

He showed a picture.

Bartender: Si, ahi.

To Lt. Dago's surprise, he saw a guy in his late twenties or early thirties dancing like crazy; he was lifting his arms like an eagle, wiggling his body and his hands, and not moving his legs from the floor, only his waist and arms. He was surrounded by three pretty girls who were dancing to the rhythm of the band, and Lt. Dago was in shock. He couldn't believe what his eyes were seeing. Then a waitress came over to him.

Waitress: Necesita algo para tomar? (Do you need a drink?)

Lt. Dago: Que me recomendas? (What do you have in mind?)

Waitress: Le traere algo… (I'll bring you something…)

Lt. Dago: Esta bien. (Okay.)

The waitress left and Lt. Dago stared to dance like the others. He tried to reach Alexandrus, but the girls didn't let him pass. They were very excited dancing with Alex.

Lt. Dago: Disculpen, necesito hablar con esa persona. (Excuse me, I need to talk to that guy.)

One of the girls: Que? (What?)

Lt. Dago: Quiero hablar con el. (I want to talk to him.)

Girl: No te escucho. (I can't hear you.)

Lt. Dago: Disculpen necesito hablar con el.

One girl: Bailemos. (Let's dance.)

And the song was over. Lt. Dago was ready to reach Alexandrus.

Lt. Dago: Disculpe, necesito hablar con usted.

Alexandrus: Hey, how are you?

Lt. Dago: You speak English.

Alexandrus: Yes and I speak three more languages.

Lt. Dago: Good.

Alexandrus: I learned from my parents.

Lt. Dago: Okay, I don't know how to begin this but…

Alexandrus: Don't worry. I know why you're here, my parents already told me, but listen, I'm not sure if I can do it.

Lt. Dago: What do you mean?

Alexandrus: I think you have the wrong guy. What makes you think that I'm the chosen one? You guys have the best army, the marines. I'm just a civilian.

Lt. Dago: Listen, I know what you've been through, but the thing is, my people need you, the entire world is in a panic, and we are preparing to launch an attack on those sons of bad b… We never thought it would come to this; we know the situation and we'll handle this attack, but if we don't act quickly, we're facing annihilation of our planet and you guys are the only ones.

Alexandrus: What do you mean…us?

Lt. Dago: Yes. See these pictures?

Lt. Dago pulled two pictures out of his pocket, one of a guy the same age as Alexandrus, and a beautiful girl with black hair, an athletic body, and black eyes.

Alexandrus: She is beautiful, and who are these guys?

Lt. Dago: Well, these kids are the future of this planet like, you know—your parents Ingui and Une were the survivors from the crashed plane. They and Currumiche were the three aliens inside of that ship. They were captured by our government, but somebody helped them to escape, and you know the rest.

Alexandrus: Yes, my parents told me all about it, but I never thought that some day it would all come to this. Your government was after my parents.

Lt. Dago: I hope it's not too late. Please, my superiors are expecting us.

Alexandrus: How do I know that you guys are telling me the truth?

Lt. Dago: Please, this time is different. The government of the US needs you.

Alexandrus: I'll do it, not for your people, but for the people on Earth.

Lt. Dago: It's okay, I understand the situation about your parents.

Alexandrus: No, you don't have any idea about that. I've got something to do first.

In fact, he had a big game the next day; it was a getaway for those people who suffered from the war. Lt Dago didn't have a choice; he knew that people from Earth were living a nightmare from the Zurdos attack and he understood also that by letting Alexandrus play, he was giving them some faith and hope.

Lt. Dago went to the local stadium the next day to see this guy. The people were excited, and only one guy was standing in the game, not like Alexandrus but his skills were comparable to Alexandrus's skills. He was Atlacatl (a local warrior that fought against the Spaniards) and he was really good. He was one of Alexandrus's best friends. After the game was over the two friends hugged each other. They were preparing to leave as soon as possible.

Lt. Dago: Nice game, guys, but we don't have too much time; we must leave now. They are waiting for us.

Atlacatl: Sir, I want to go too. I grew up in the air force and I fought in Iraq.

Alexandrus: We might need him, sir; besides, he's the only one that I trust.

Lt. Dago: Hmmh. I don't know, but…okay, you're in.

Alexandrus: Atlacatl has his own platoon.

Lt. Dago: What?

Atlacatl: They're the best, they also fought in Iraq, and they want to defend our planet.

Lt. Dago: Okay, take them—we'll need as many as we can get. We're moving on.

They left. Alexandrus and Atlacatl's platoon were around 15 soldiers.

Lt. Dago: Okay, soldiers, you know the situation and all the risks.

Atlacatl: Yes sir, Alexandrus's parents told us everything that would happen.

Lt. Dago: Yeah, but you don't know the enemy; this enemy is stronger and they don't have mercy. They're killers and they want our planet for themselves.

Alexandrus: Well, where are the rest of your people?

Lt. Dago: Patience, soldiers, we're going to meet them in a few hours.

They were escorted in a plane. In the plane everybody was a little nervous because of the magnitude of this event. Alexandrus was really close to the window, and he was watching the landscape of the mountains from ESA. He was thinking that this trip might not have a way back, but he had to go—he had to do it for the planet Earth.

Alexandrus grew up watching and playing soccer. He was a local hero, and he was very good with the ball. He was also thinking of his family and the land where he grew up.

Lt. Dago: What are you thinking?

Alexandrus: It's not going to end the world like this?

Lt. Dago: I hope not.

Alexandrus: I won't let the Zurdos take our home even if it's the last thing in my life. They're preparing for the final assault.

Lt. Dago: No, we won't let this happen.

Alexandrus: Tell me, Lt. Dago, how in the world are we going to make it to Europa?

Lt. Dago: Don't worry, Alexandrus, everything is under control.

Alexandrus: I hope that you're right because my parents think different.

Lt. Dago: How about you, Atlacatl? You've been quiet the whole trip.

Atlacatl: Sir, I wanted to join forces with you guys when I found out the whole thing.

Lt. Dago: Why?

Atlacatl: Simple, sir, I have a family to protect.

Lt. Dago: That's a good answer.

Alexandrus: Sir, my friend can be very helpful, believe me. We've been working out since we were kids, and this guy is a strong warrior; his men are the best in the area. They can survive in extremely hostile environments.

Atlacatl: Listen, sir, I just need a chance. We need to protect our planet from the Zurdos. I want to see my kids grow up with a future.

Lt. Dago: Welcome aboard, our planet is waiting for you.

After a couple of hours Lt. Dago, Alexandrus, Atlacatl, and the rest of the group landed in Houston at a big base. The construction was fascinating. They got off the plane and were escorted into another facility.

Commanding Officer: Welcome aboard, sons.

Lt. Dago: Thank you, sir.

They saluted in the army fashion.

C. Officer: You must be Alexandrus.

Alexandrus: Yes sir.

C. Officer: And you?

He turned his head to Atlacatl, who was really surprised by this place; he had not seen anything like this before.

Lt. Dago: Let me explain, sir, these soldiers are amazing.

Atlacatl: My name is Atlacatl sir, and I'm at your service.

Lt. Dago: Sir, I can explain this.

C. Officer: It's okay, Lt. Dago, we need brave kids like you. You are the future…all of you. As you know, everybody hopes that you guys are going to defeat this enemy called Zurdos, and I'm sorry about your parents.

Alexandrus: It's okay, it wasn't your call but we'll do our best, sir.

C. Officer: Come on, people, I want to introduce you to somebody else, follow me.

They left the room and were escorted into another big room for meetings, where they found two more people waiting in the stairs: Amadeus and Dulcinea.

C. Officer: Let me introduce you guys, Amadeus and Dulcinea.

They stood up and they were very courteous.

Dulcinea: You must be Alexandrus.

She was the first one to stand up.

Alexandrus: Yes and you must be Dulcinea.

He turned his head to Amadeus.

Alexandrus: And you are Amadeus?

Amadeus: Yes, we already have heard a lot about you.

Alexandrus: Well, I'm honored, but I also have heard a lot about you guys.

C. Officer: Sit down, please.

Two high-profile officers joined the meeting, Colonel Demello and Captain J.J. Colonel Demello was a huge, tough black guy with many medals of honor. He was a dignified army man. He was going to be in charge in the mission to Europa. Captain J.J. was also a tough guy, but in contrast with Colonel Demello he had no mercy; he was blamed for killing thousands of people in the war from North Korea. He was in command of the US troops in Asia, and he was a mean guy. He did not care about the civilians from the other countries and no-body dared stand against him. He was a loyal guy in the army, and it was a good reason to bring him to this mission.

C. Officer: Let me introduce you. Colonel Demello, who is in charge of this mission, and Captain J.J.——he is second-in-command.

After a brief period they began discussing something that the new guys could not hear.

Amadeus: Excuse us, but we haven't heard anything about the mission yet.

Colonel Demello: Okay, kids, let me be clear with all of you, I didn't like the idea that you were brought here. We're at war and I don't know who you are. I've been told this is a serious mission. We're sending our best people, the best of the best in the world and with the permission of the Commanding Officer, I don't know if you can handle this. Do you have any idea of the magnitude of this?

C. Officer: Please, Colonel, calm down, we need these kids. They can make the difference. They're half humans, half aliens. They might know something.

Captain J.J. was listening to the conversation. He didn't like the idea of going with these new people either.

Captain J.J.: Excuse me, sir, but I agree with Colonel Demello. We don't need these "children" for this. We have recruited the best of the best in our army and it's not just that we have the best of the best in the world. It's that we're not going to waste our time babysitting these kids.

C. Officer: Captain!

And that's when our heroes began to surprise everybody. Since the moment they were introduced, they started to communicate through telepathy, which they inherited from their parents. In that room the officers were discussing how to deal with their differences, but our heroes were having fun.

Alexandrus: Hey, it's nice meeting you guys.

Amadeus: It can't be real, my father told me about this before, that this attack would happen, but I thought he was kidding

Dulcinea: Yeah, when I was a little girl, they taught me a lot of strange things: languages, politics, psychology, and even how to defend myself. I'm glad that they were very strict with me. They told me of this day, and I didn't believe them, but now I think it is true.

Amadeus: Hey, they told me the same, that someday we were going to save this world. I thought they were kidding. They're expecting a lot from us

Dulcinea: Our parents didn't finish their quest.

Alexandrus: I think is our duty to defend those people.

Amadeus: Why us? Don't get me wrong, but we have the marines, air force, and the army; they're more qualified soldiers than us.

Dulcinea: Yes, you're right but they can't do it alone. I think we might play an important role over there. I'm not sure yet how, but...

Alexandrus: Hey, my dad loved this planet, and thanks to my mom, who was working as a member from the cleaning crew in that place, she could save him and your parents too. I have to do this. My dad trained me so hard with fighting skills; it was difficult at the beginning, but I didn't have a choice. I don't know if it's going to help me in this quest, but we should be glad to be here. We're going to stop these Zurdos at any price—that's what counts. There are good people waiting for us, remember that. Our parents would feel very proud of us.

Dulcinea: You're right. My dad told me that we have to free them from the Zurdos.

Amadeus: I hope those people are still alive.

C. Officer: Excuse me, excuse me, are you guys here?

Alexandrus: Sorry, we were talking among ourselves.

C. Officer: That's what I was explaining to Colonel Demello and Captain J.J., about your skills. With your powers we can beat our enemies.

Captain J.J.: Yeah, right, what power? They're a bunch of guys looking for 15 minutes of fame. We're in a war and we don't even have a chance against them. We need soldiers, warriors willing to defend our planet, not psychologists.

It was at that moment when our heroes showed their skills.

Dulcinea: I know what you're thinking, Captain, you don't have tomorrow. What you did in Asia is killing you! You don't sleep and one more thing, please stop watching my body—you're not my type.

Amadeus: You don't feel comfortable with us because you don't feel comfortable with yourself; you're jealous of Colonel Demello.

Captain J.J.: What that f… are you talking about?

Amadeus: And you, Colonel, you're scared to go out in this mission; you're afraid that you're not going to come back, that you'll lose your investments in Wall Street, and you're also suspecting that your wife is cheating on you.

Colonel Demello: What's that?

C. Officer: I told you guys they're special. I knew their parents. They were a peaceful people, and we'll not make the same mistake twice, and that's it.

Colonel Demello: I hope that you're right, sir; I hope it's not a big mistake.

C. Officer: Okay, people, in 21 hours we're leaving and God bless you all. You're dismissed and rest while you can.

They were then taken to their rooms.

Alexandrus: See you tomorrow, guys.

Dulcinea: Get some rest. We'll need it.

Amadeus. Okay, you too, guys.

That was a bad start between the officers and the civilians but they had to get along. They were escorted to their respective rooms, but nobody could sleep; each individual was too impatient to sleep. Colonel Demello, in his room, was remembering the conversations with the new guys. Captain J.J. was drinking in his bedroom, trying to forget the horrible things that he did in Asia. Dulcinea was remembering all the things that her parents taught her—to always be a good person, to be alert, to be patient with the right guy, to be smart, to be good, on the right path. Amadeus was thinking too. He was very intelligent and he was always looking for women but nothing serious. But in this case he was very serious about Earth, and he wanted to be there, Atlacatl was thinking of his family and his children, and finally Alexandrus was thinking of his family, his girlfriend, and most of all, he was seeing the planet Earth living in peace again. He wanted to fight the enemy. He had seen a lot of destruction in the world, and he wanted to change it. He wanted to give it a chance, the planet where he grew up, and he wanted to see the people again smiling. He wanted to see a future.

20 hours later.

After they brief meetings almost everywhere, finally they were ready. Everybody had a military uniform and they were heading to another facility, where they saw more people, more equipment. They were guided to a much bigger platform: the naval base "Scorpion," which was a big base in which the US government built two spaceships, the "Bush Pride" (honoring former president Bush, who started the program) and the "Sand Whale," the best and biggest spaceship in the world. This war spaceship had room to carry sophisticated technology weapons, radar supplies, and food for at least two years. Also the UN was sending another two spaceships, the "Chameleon" and the

"Earth" (from Russia), both of which were settled in Europe. They were very sophisticated too. With these four big spaceships the planet Earth was hoping to defeat the enemy.

Alexandrus: Oh my God, this is incredible.

Dulcinea: Wow! This place is unbelievable.

Amadeus: I never thought I would be in a place like this.

C. Officer: This is just the beginning. Take a look to your right.

Everybody was really surprised by what they were witnessing. They saw the two big ships, built over the ocean.

Atlacatl: So this is it, we're going to fight for this planet.

Dulcinea: And how long is it going to take us?

Colonel Demello: Approximately two weeks.

Amadeus: Two weeks?

C. Officer: Yeah, it's going to take you guys around two weeks. With this technology and the "Super Eurotanius" we have developed such machinery.

Alexandrus: "Super Eurotanius"? What's that?

Dulcinea: Didn't your parents tell you that? It's supposed to be the best and strongest mineral of this galaxy. With the Super Eurotanius any spaceship can reach maximum velocities in space—of course, only when this mineral is in a liquid stage. This was top secret in this government.

Atlacatl: And what happens when it is found in a solid stage?

Amadeus: You can make a lot of constructions. This mineral is not heavy; it's a light mineral and it's very resistant to anything.

Alexandrus: Wait a minute, why didn't my parents tell me that?

Dulcinea: You were very busy playing soccer, ja, ja.

Amadeus: But this is not all.

Alexandrus: What?

Amadeus: We're going to Europa not only to fight the Zurdos.

Atlacatl: What do you mean?

Amadeus: We're going to find this Super Eurotanius and bring it here, am I right? Our parents were chased for that.

C. Officer: You're right, it's like a petroleum replacement.

Colonel Demello: Do you have any idea what Eurotanius can do for mankind?

Dulcinea: Are we going to steal it? That's unfair. They have hope in us.

Colonel Demello: No, we're just going to borrow it.

Dulcinea: I can't believe it, there's another group of good people waiting to be saved and we're just going to go for this kind of petroleum.

Alexandrus: Sir, with all due respect, I think Dulcinea is right.

Captain J.J.: You don't have to worry about the people on Europa—they're dead.

All heads turned to the Commanding Officer.

Alexandrus: Sir, is it true?

C. Officer: Yes, it's true, the Zurdos killed the rebels, all of them.

Amadeus: But it can't be, there were good people living over there.

C. Officer: I'm sorry but we need this Super Eurotanius for ourselves; we used it all in the ships' construction and we need more of it. We can't allow the Zurdos to have it on their hands.

Suddenly a siren went off.

Alexandrus: What's that?

C. Officer: It's time to go. This is our big moment.

They were escorted right away to the Bush Pride, and there was another group of people getting inside the Desert Whale. In the Bush Pride were our heroes with approximately 200 men. In the Desert Whale were about 3,000 people. The UN space ships held approximately 2,000 men each; the Chameleon was built in England

and the Earth in Russia.

(In the entrance of the Bush Pride)

Alexandrus: Okay, guys, let's go.

Atlacatl: Let's go and God bless us all.

Amadeus: I don't like this mission.

Dulcinea: Neither do I, but we don't have a choice; we have to protect this planet too. I know there's people alive over there.

Amadeus: Even if you're right, we're going to free them from slavery—then what? Are we going to steal all their technology? Our parents chose the wrong alliance.

Alexandrus: Relax, guys, I don't like this either, but they're right. First of all we have to take care of ourselves. That means our planet and then we'll see. Remember, don't speak of this again, especially in front of Captain J.J. We know that he hates us, so don't give him another reason.

Dulcinea: You're right, let's go.

Colonel Demello: Okay, everybody, move, move, move, hurry up. This is our chance; the sun is really away from our course. Move, move to your positions.

Everybody was running inside of both ships. They were really ready. It was like a dream to them, being a part of history. It was the first time that mankind had embarked on such a quest, and they knew that they probably wouldn't even make it to Europa. But they had to take the risk. Embarking on a new odyssey into space, the first humans crossing our galaxy, the first ones to go farthest from our moon, the first ones to go beyond our imagination. There were some screens on the other side of the ocean where the people could watch the names and faces of each soldiers. Those people were witnessing the launching of our heroes, and it was a very emotional moment. The takeoff was very uncertain but they felt proud and patriotic.

From the speaker's tower: Nine, eight, seven, six, five, four, three, two, one, zero—ready…and God bless you!

There were millions of people watching them take off, including the most prominent leaders of the world, and they were so nervous until the spaceships took off in orbit. They screamed so loud, the people made a lot of signs, some people were crying, some soldiers were leaving their children, wondering if they were going to see them again, and there was sadness and happiness at the same time. The people left on Earth were really proud of them.

Now the ships were in orbit, heading into a new journey in the sky. In the Bush Pride were our heroes, but not all of them were together.

On one side of the ship were Amadeus and Alexandrus, and in the other part of the spaceship were Atlacatl and Dulcinea with the rest of the troops. They were still in tears.

Alexandrus (to himself): Wow, what beautiful scenery from this point. The Earth is so big. How many people are hoping that our troops make them pay? It's going to be a really hard task but God is going to bless us all. It's a privilege to be here and I'm not going to let them down.

The first step was done; they were crossing into space, and they were experiencing turbulence. Everybody was still in their seats, and they were traveling so fast. After a couple of minutes they were getting used to space.

Amadeus: Hey, Alexandrus, did you notice that everybody is still watching the Earth?

Alexandrus: Yeah, the Earth is below us and look at them, they're glad to be here.

Amadeus: Like us, right?

Alexandrus: Yes, like us.

Amadeus: Do you think we'll make it back?

Alexandrus: I'm not sure, not all of us will make it, but at least whoever does, the people of Earth will be glad. We're doing this for future generations. I want to see my family again, I want to see the smile of every single human.

Amadeus: Me too, I want to see my family and friends.

Alexandrus: I made a promise to myself that I will come back alive to my family.

Amadeus: I think I tried too much, but now I just want to end this war. There's a lot of innocent dead people and I think this is the only way to protect us, going over there. I want to be part of something.

Alexandrus: What did your parents tell you about Europa?

Amadeus: Well, before they decided to escape from their planet, they told me they used to have a great culture and beautiful cities;

they were very advanced if you compare it with planet Earth, and they were ahead of us about 20 to 30 years. They said that our planet is heading towards that destination.

Alexandrus: Yeah. My parents told me that it all started when they were doing experiments with DNA from humans and DNA from animals. My father was a doctor over there.

Amadeus: And not only that, they used or mixed it with insects and plants.

Alexandrus: Yeah, they made a big mess, and our parents refused to make such atrocities. They were nuts.

Amadeus: Aha, and it's when they decided to leave their planet and came to Earth to warn the people, they barely made it.

Alexandrus: But they got in the wrong hands, and they stole the Super Eurotanius.

Amadeus: Yeah, but like in Europa, there's still a lot of good people.

Alexandrus: You're right, there were good people on Earth too. Without my mom and the help of Professor Darwin they would never have escaped from the government; they were very lucky, not like the ones in Area 51.

Amadeus: But I can't believe the way they were described like green creatures.

Alexandrus: You're right. I think it's our destiny to be here, and it's a pleasure to be with you and the rest of these people.

Amadeus: The pleasure is mine and I have faith in God that we're going to beat them.

Alexandrus: But how has Europa changed since?

Amadeus: I don't have any idea.

Alexandrus: Do you think it's too late?

Amadeus: I don't know. I hope not.

Alexandrus: The atmosphere was the same as Earth's when our parents left it.

Amadeus: Hmm, but I don't think it's going to be like our parents told us.

Alexandrus: I hope everything is going to be all right.

Amadeus: And the Super Eurotanius, what do you think about it?

Alexandrus: It's another reason that we're going to invade Europa. It's true that our planet is running low on gasoline and we don't have any reserves, but they're right, we have to destroy the Zurdos because they can create more bombs and they are going to use them against us, and a second assault our planet is not going to handle. Mankind will disappear with all its history.

Amadeus: How are we going to win this war? The Zurdos will know right away that we're heading there right now.

At that moment Captain J.J. came up.

Captain J.J. (in an ironic tone): Excuse me, kids, having fun watching the planet Earth?

Alexandrus: Yeah, we should come more often. Next time I'll bring my soccer team for celebrations.

Captain J.J.: Very funny, stupid…

He grabbed Alexandrus by his shirt, and Amadeus came to help Alexandrus.

Amadeus: Calm down, Alexandrus, he's our boss. Don't do anything…yet.

Colonel Demello (approaching): What's happening, Captain?

Captain: Nothing serious, I was just telling them that you need to speak to them in our conference room.

Colonel Demello: Yes, report at the office in ten minutes.

Alexandrus and Amadeus: Yes sir, we'll be there.

They went back to their positions, and Captain J.J. left, really upset about what had happened.

A voice sounded across the ship: Will the following please report to the conference room in ten minutes: Alexandrus Viscarret,

Dulcinea Serrano , Amadeus Eser, Atlacatl Martinez, James Gordon, and Robert Birkate.

Colonel Demello was waiting to give some details about the mission. Alexandrus and Amadeus headed to the conference room, and there they found the rest of the crew waiting at a big round table with a big screen hanging on the wall. Then they stood up.

Colonel Demello: Thank you for coming in. I need to be clear; we all are in this together. We're on a special mission and there's probably no way back home, but we'll do our best to defend our planet...our families...our generations...our future. We were chosen for this and I want you all to be the best. We have to locate the City of Fury; we don't know exactly the location, but our system indicates it could be north. However, we're heading south and, as you all know, this is the place where our enemy is standing. They have broken the line and they have the guts to attack our planet. Here on this screen, we have data collected about this place. We're not going to let anybody take us away from what is ours, ladies and gentlemen. We're here because we're standing as united nations and we need to free those people if they're alive. I know it's going to be hard, but if we stick together we can do it. The Zurdos have total control of the moon; they had enslaved the people of Europa. Once they had big dreams like any flourishing civilization, but now they might be the last of their kind. The Zurdos didn't want to share the moon with them, so they decided to execute them, and now they want planet Earth. They don't want us to be alive. They are a race of hunters and we will be their trophies. Your mission, people, is to destroy the Zurdos and try to locate any survivors.

Lt. Dago: Excuse me, sir, but how are we going to attack them and how are we going to get there, once we land?

Colonel Demello: Good question. We will launch some missiles from here and we will wait. We have our backs covered by the Desert Whale ship; it's the biggest machine ever built by human engineers. We have the weapons and technology to defeat the Zurdos and we're also escorted in this mission by Earth and the Chameleon, the two greatest war spaceships ever made, and once we put our feet on the ground, we have sophisticated mask weapons capable of controlling radiation and biological and environmental conditions on the moon.

Dulcinea: Do we know the conditions of the planets? We probably don't know exactly the location of the City of Fury. North or south? It's not enough. Do we know if there are people still alive from the Zurdos? If there are, can we trust them? Do we have complete information about this mission?

The people are murmuring.

Alexandrus: (to himself) *I like this girl*. Yeah, those are great questions.

Colonel Demello: We know a little about the conditions of Europa, based on our data. We should find a lot of fog, rain, and the place is like our planet. Since the past governments discovered life in this place, they kept it in secret. The location is uncertain, but we know that they will find us and about the people. Well, I want to be honest with all of you people, the last time that we heard about them was a long time ago, thanks to these guys' parents (he pointed at Alexandrus, Dulcinea, and Amadeus) and that was because we ignored the threat from the Zurdos. From that time, everything has changed. We don't know if they are still alive, we don't know what the Zurdos might look like after what they did in their lab experiments, we don't know anything. We just want to destroy them. Any questions?.

It was in that moment of the colonel's speech that someone hidden in the group showed up—Professor Kishimoto.

Professor Kishimoto: Let me add more to that question.

Captain J.J.: Who the heck are you?

Professor Kishimoto: Hmm, let me introduce myself. My name is Joseph Kishimoto, and I'm an archeologist with a PhD in astrophysics. I have a degree in seven languages and I'm also your counseling man.

Alexandrus: Very well, I'm glad to have a person like you aboard.

Professor: Thank you. I'm honored, but my mission here is to help each one of you. This place is covered with ice at 270 Fahrenheit below zero on the primary external surface. There's no way that any human could've lived at such temperatures in the past, but we all know about the icy spectrum that they created for hiding themselves. It's the size of nine-tenth of our moon, from what we discovered in the past. Europa has an ocean underneath, and that ocean created an atmosphere capable of sustaining life, and as we suspected, there's a heat source at the moon's core that keeps it warm too. Once we're in its orbit, we'll encounter the intense radiation that surrounds Jupiter, so if we're going in, the big red spot in Jupiter is like a black hole. It could swallow our ship into a big crash; it's like a big magnet, so we better be careful. Once we target the site, we're going to require a nuclear power device which is a big virtual torpedo. It will create a hole and it will hold it for a couple of minutes, maximum one hour, and then it will freeze up, so after we make the hole, we must get inside as fast as we can. We don't know what will happen after, but we should be fine to get back.

Colonel Demello: Well, that is better explained. If you will excuse me, you all are dismissed. We'll be landing soon, so if I were you guys I would rest because who knows when we are going to get another chance.

Alexandrus: I don't get it. Why? I think they're not telling us everything.

Dulcinea: In front of everybody? Come on, Alexandrus.

Amadeus: Dulcinea is right, he's just receiving orders from above.

Atlacatl: But one thing is for sure.

Amadeus: What?

Atlacatl: We should stick together in this thing. This is the only chance that we're going to have, and I want to tell you guys one thing. I want to come back to Earth, I want to see my family again, and I want to see the freedom we once had.

Alexandrus put his hand on Atlacatl's shoulder.

Alexandrus: You're right, my friend, I want to come back to Earth. I want to see my family too, and I miss those soccer games, especially against you.

EUROPA

Amadeus and Dulcinea put their hands on each other.

Dulcinea: We all agree on this. We should stick together and maybe it's going to be the only chance that we have.

Alexandrus: Okay, guys, what do you know about Europa?

Dulcinea: Nothing much.

Amadeus: The same that we were taught since we were kids.

Alexandrus: Yeah, you're right, we should follow the colonel's advice and take a rest because who knows.

Atlacatl: You're right. See you guys later.

Amadeus: See you in a bit.

After the meeting everybody went to their respective rooms; nobody rested but Amadeus, who was sleeping with his mouth open and drooling.

It had been several days since they took off (the universe was infinite) when all of sudden a big noise broke the silence. It was the alarm from the ship and a louder voice on the speakers.

Speakers: All personnel, please go to your respective positions. We are getting close to Jupiter's atmosphere; we're getting close to Jupiter's atmosphere.

Alexandrus (In his room): Oh my God, we're there, I've got to go to the cabin. (Telepathically to Dulcinea and Amadeus) Hey guys, are you awake?

Dulcinea: Yeah, I'm awake. I felt something weird.

Alexandrus: Yeah, I felt it too. Don't go yet…wait for me.

Dulcinea: Okay, but hurry.

Alexandrus went right away to Dulcinea's room. He was watching people running very scared, and the siren was still on.

Alexandrus: Are you ready?

Dulcinea: I'm scared but I'll do it for my parents and for the people on Earth.

Alexandrus: Me too, I'm ready.

Then they heard Atlacatl.

Atlacatl: I lived for this too, and you're the only ones that I could trust in this mission. There's something else that they're not telling us, but I'll stick with you guys. I've got my men ready too.

Amadeus: You're not thinking that all of you guys are going to do it by yourself; you're going to need extra help.

Dulcinea: Of course, you're welcome too; together we can make a difference, and from now on, you guys are my family.

Alexandrus: Yeah, from now on we're family.

Atlacatl: Hey guys, I think we should go. Everybody is in the cabin.

Amadeus: Yeah, let's move on. We're going to be late.

Everybody left the room, heading to the main cabin, where the big commanders were standing. In the cabin they were in shock; they couldn't believe what their eyes were witnessing. The Russian ship Earth was falling down into the red spot in Jupiter, and they could not do anything. They were listening to the last signal from the spaceship—they were screaming, asking for help, and nobody, nobody did anything.

Alexandrus: There's nothing we can do about it.

Dulcinea: Come on, guys, do something. There are humans, and they need our help. We're not going to leave them like that. They can use the centrifugal force.

Amadeus: Alexandrus is right, Dulcinea, there's nothing we can do. It's too late.

Earth ship voice: Please, help us, we're going down! Do something, please!

The Earth ship was falling quickly into the red spot; it was attracted right away to the magnetic field from it. Then they heard a big noise.

Colonel Demello: I'm sorry, guys, we can't do anything... God bless you all.

Colonel Demello almost cried when he said that, and most of the crew were crying about the destruction of one of the spaceships.

Colonel Demello: Ladies and gentlemen, with all due respect, they deserved much better, but we are all in the same predicament. Now we're going to finish this mission without the help of our Russians brothers; we're in this quest together and we'll finish it together.

Guy #1: Sir, we're leaving Jupiter's magnetic field and we're closing in to Europa's orbit. In a couple of minutes we'll be landing.

Colonel Demello: Okay, everybody listen up. We don't know what is going to happen over there, but I'll tell you one thing: people on Earth have put their hopes on us, and we're not going to let them down, so be ready and so be it.

They barely manage to avoid the red spot, and the whole group starts to cheer.

Everybody: Yes, for our brothers on the Earth ship…for our generations… Yeah.

They felt very patriotic moments after what they saw with their Russian brothers. While still in shock, they went to their respective positions.

Guy #2: Sir, we're getting closer.

Colonel Demello: Okay, son, get us there.

All the nervousness came up once again—there was nothing safe in this quest—and the seconds seemed like hours, but finally they crashed into Europa's orbit.

Guy #1: We're in Europa's orbit, sir. It seems inhabitable, but…

Colonel Demello: What the heck happened here?

Guy #1: Nothing here seems to be what it seemed back on Earth.

Indeed, the illusion of a deserted, icy moon was wrong; it was like a gigantic biodome covered with ice. Meanwhile, our heroes were discussing the orbit.

Alexandrus: I don't like their plan.

Dulcinea: Me neither. There's something wrong.

Alexandrus: We better tell Professor Kishimoto about this.

Amadeus: Guys, I don't get it. Correct me if I'm wrong, but we're talking about the landing, right?

Dulcinea: Of course. We're not going to have that much time to dig, and then to get in, we're out of time.

Alexandrus: Yep, we don't have an hour. Jupiter's radiation will fry us alive.

Dulcinea: I would say if we're lucky, we'll have two minutes.

Atlacatl: Two minutes? That's impossible.

Alexandrus: I know. With that time we'll be lucky if we just pass one.

Dulcinea: We better tell Professor.

Alexandrus: We better hurry up.

The professor showed up. There was only one spotting landing on the other side of the moon, but Jupiter was so close that they couldn't find the perfect timing.

Professor: I know what you're afraid of.

Dulcinea: Professor, I think you're wrong in your theory on how to get in.

Alexandrus: Yeah, she's right. We can't prove it, but there's something wrong with the surface. The spotting landing can be risky.

Professor: Don't worry, I might be wrong, but this is a great risk that we're willing to take.

Dulcinea: We have to dig all three ships at the same time.

Colonel Demello (passing by): Everything is under control; each ship has the nuclear reactor to dig, but we need to save fuel.

Dulcinea: You don't think that's unfair? What about the others— have you asked them to wait? Assuming that we don't have enough power, they'll be burned.

Colonel Demello: We'll dig. They have to wait; they are our back-up. We are supposed to follow orders.

Alexandrus: I guess he's right in this, Dulcinea.

Amadeus: I agree with this too.

Colonel Demello: The Desert Whale is our big chance to survive, and our brothers from Chameleon are going to support us.

Suddenly they landed on the surface. It was very cold, and they could see Jupiter and its other moons. They spotted the site where they were supposed to dig in.

Guy #2: Sir, we're in, we got it.

Colonel Demello: Okay, let's make history.

Time wasn't on their side, and it was hard to dig in. They were facing the red spot again. The Bush Pride began to dig, but they found a big obstacle.

Guy #1: Sir, we have a problem.

Colonel Demello: What?

Guy #1: We don't have enough power to continue. We used it all. We're out of fuel. When we were trying to escape from Jupiter's atmosphere, it left us out of it, and we can't continue doing this.

Colonel Demello: Okay, guys, put me in contact with the Desert Whale.

Guy #2: Okay, sir, you're in.

Colonel: This is Colonel Demello. I'm requesting that you finish what we came to do. We're out of fuel, repeat, we're out of fuel.

Desert Whale: Understood, sir, we'll finish it.

The Desert Whale began to dig at the same spot, but in that time the people did not realize that the radiation of Jupiter was beginning to spread.

Guy #1: Sir, we have another problem.

Colonel Demello: Now what?

Guys #2: We've been exposed to Jupiter's radiation.

Dulcinea: Yeah, I'm feeling that too; we better hurry.

Amadeus: Me too, it's getting hot.

Colonel Demello: Hurry up. The radiation is getting us.

Alexandrus: And not only that, sir, in a couple of minutes we're going to face the red spot again, and I don't think this time we're going to be lucky.

Professor: You're right—another encounter with the red spot and we're history.

Colonel Demello: Hurry up, boys.

Guy #1: Sir, Captain J.J. is wondering why it's so hot.

Captain J.J. was in another level from the Bush Pride in a different cabin.

Captain J.J.: Sir, why is it hot? Our troops are getting impatient.

Colonel Demello: Tell our boys it's just a matter of time. We'll get them out of there right away.

Racing against time, finally the Desert Whale made a hole on the top of the dome, but the misfortune happened again to our heroes; drilling a hole meant they were exposed to any threat.

Desert Whale: Sir, sir, we're in trouble. Our ship is not responding, our ship is not responding, we're going down, sir, we're going down.

In fact, the Desert Whale made the hole, but it was too much exposure to the radiation of Jupiter that made it burn quickly.

Alexandrus: There it is, sir, let's go; we don't have enough time to get in.

Dulcinea: Yes, yes, please, we don't have time. We just have a minute.

Amadeus: You don't believe us, ask Professor Kishimoto.

Kishimoto: They're right, sir, we don't have too much time. Let's go…now!

The Desert Whale entered the subsurface from Europa, burning until it crashed into the ocean; the Bush Pride went after it, followed by the Chameleon. When Chameleon was entering the hole, inexplicably the hole made by the Desert Whale started to freeze up, and when that happened, Chameleon was thrown into another place on the subsurface of Europa; the Bush Pride was sent into another direction and landed in a place like an ocean or a big lake. From the velocity that they were sent from the surface from Europa, most of the crew were unconscious.

The spaceship luckily survived the crash against the water, and after a couple of minutes, there were some voices.

Alexandrus: Is everybody okay? Hey, are you guys okay?

He looked around. There were people lying on the ground, and then he heard somebody else.

Colonel Demello: I'm okay. Go get the others.

Amadeus: I'm here, I'm okay. Dulcinea, are you all right?

Atlacatl: Hey guys, she's here. Let me help you.

There was a big cabinet on her foot.

Dulcinea: Thank you, Atlacatl. I think we're okay.

After everybody recovered from the crash, they looked right in front of the screen. In that spot there were Colonel Demello and Alexandrus. They were the first ones to be standing in the screen of the ship. They were astonished one by one, Dulcinea, Amadeus, then Atlacatl. They got closer to Alexandrus and Colonel Demello, and some members of the crew were trying to get up.

Dulcinea: What's going on? Oh my God.

Atlacatl: What's going on out there?

When Atlacatl came up to them, he witnessed what the others

were amazed by, and effectively what they were seeing were endless shores like the beaches of Earth. There were palm trees, and the environment seemed like a great summer vacation in California, with the difference being that it was empty. But the weirdest thing was the sky. On Earth, the sky is an imminent, deep blue "ocean" surrounded by the greatness of white from the clouds; there it was covered by a big roof of grey. It seemed to be cloudy all the time, but you could see the giant red planet. It was a kind of magnetic shield covering the sky.

Alexandrus: Yeah, I can't believe what we're seeing.

Amadeus: It looks like Earth.

Alexandrus: It has beaches, trees.

Dulcinea: It's beautiful. No wonder we've never seen it from our satellites.

Professor: It's amazing. We never thought to imagine a place like this; somehow they manage to be hidden all the time, and they made us think this was a desert moon.

Colonel Demello: Okay, people, that's enough about the "prettiness" of this place; let's do what we came for, Sergeant.

Sergeant: Yes, sir.

Colonel Demello: I want you to send a probe outside to see what's out there, and I want you to tell Captain J.J. to be prepared for my call. Let's see what's out there.

Sergeant: Yes, sir.

Dulcinea: Excuse me, sir, what about the Chameleon?

Colonel Demello: Sergeant, I need you to make contact with the Chameleon to see if they're okay.

Professor: Or if they're dead like the others.

Alexandrus: Relax Prof, they might be okay.

Amadeus: I hope you're right because they were propelled very hard from above.

Meanwhile the Bush Pride was trying to make contact, and the probe that they sent in was giving them the first signal.

Professor Kishimoto: Sir, we're receiving a signal from the probe. Wow, it's amazing. I knew it, I was right, but nobody believed me.

Colonel Demello: Cut the crap, Professor, what's out there?

Professor: The trees are like the ones we have on Earth and there's no radiation.

Colonel Demello: What?

Professor: Well, I think the core of Europa keeps it warm enough for this ecosystem and has something to do with the atmosphere. It's a biodome.

Colonel Demello: But Europa doesn't have atmosphere.

Professor: Hmm, in this case, it's like a big shield built around Europa to keep it full of life inside and protected from the outside.

Sergeant: Sir, Captain J.J. is ready to send the first recognition platoon.

Colonel Demello: Send them right away

The first recognition platoon was made up of about ten masked soldiers carrying equipment. They had radio contact with the spaceship and were sent on assault boats.

Colonel Demello: Okay, boys, make history…and let me know when the transmission is ready, Sergeant.

Sergeant: Okay, sir, mm… Sir, Captain J.J. is here.

Colonel Demello (looking nervous): Bring him in.

Captain J.J.: What's going on, sir?

Colonel Demello: We're just waiting for the signal.

Time seemed to move slowly. They were getting impatient.

Sergeant: Sir, we have a signal.

Colonel Demello: Great, it's about time, put it on the screen.

Sergeant Romero: Yes, sir.

All the crew were inside the main cabin, waiting for the signal of the recognition troops.

Sergeant Romero (lying on the sand): Sir, sir, can you hear me?

Colonel Demello: Yes, son, I can see you also.

But there were some difficulties in the signal; sometimes the image was blurry.

Colonel Demello: Tell me, son, what do you see? Is it safe to go outside?

Sergeant: (static noise) Our system indicates this place has oxygen, repeat, oxygen.

Colonel Demello: What? Son…are you there?

Alexandrus: Why is there static noise here? We're too close.

Amadeus: Yes, there's something I don't like about it.

Dulcinea: Sir, I think you should bring them back.

Colonel Demello: Please, quiet. Sergeant Romero, are you there?

Sergeant Romero: (static noise, and then) Yes sir, we have some difficulties in the transmission. We have made a perimeter around the area, and what the hell is that? (static noise)

Colonel Demello: What happened? Answer me, Sergeant, what's happening?

Lt. Dago (afraid): I don't know, sir, I don't know, we just lost the signal.

Dulcinea: Sir, please bring them back, it's not safe.

Alexandrus: She's right, bring them back. They're in danger.

Amadeus: I think it's too late.

Colonel Demello: What? Answer me, son.

All of sudden they had a signal on the screen. The video camera was dropped off from the distance where they vanished. It was a desert, and there was no sign of the soldiers, no sign of any confrontation. It all happened fast.

Colonel Demello (very scared): Captain, are your men ready?

Captain J.J.: Yes, sir, do you want me to send them right now?

Colonel Demello: No, Captain, I'm going with them.

Professor: What?

Alexandrus: Sir, I don't think it's a good idea.

Colonel Demello: I want to rescue our boys.

Captain J.J.: Sir, with all due respect, I think you should stay here.

Colonel Demello: We'll see.

He turned and left the main cabin for Stage M, the cabin where their troops were waiting, but Colonel Demello was followed by our heroes and the professor. They had no idea what was happening with their soldiers.

Colonel Demello: Okay, ladies and gentlemen, what you just witnessed outside is uncertain; we don't know what's going on out there, we don't know if they are still alive, but I do know one thing for sure. We came for our people back on Earth. They have faith in us, and we're not going to let them down, and do you know why? Because we're the best of the best, so go on, and God bless you all.

Soldiers: Yes, for our future generations.

Dulcinea: Be careful.

Alexandrus: Watch your step.

The last brigade of thirty commandos were leaving Stage M in combat rafts. They were carrying masks and weapons too.

Sergeant Martinez: Okay, ladies, you heard the boss. We came here for one thing, to defeat the Zurdos.

The wind was strong, and their boats were going quickly into action.

Everybody: Yes!

Some of them were looking at the pictures from their families. Finally they got to the shore, and they started to run faster and faster, looking for their friends.

Sergeant Martinez: This is Sergeant Martinez, can anybody hear me...over?

Colonel Demello: I hear you well, son. Give me your coordinates and tell me if you see any trace of our recognition platoon.

Sergeant Martinez: Sir, we're here in the (static noise)...some place that the recognition platoon was (static noise)... I see the camera, but I don't see them.

Alexandrus: Sir, tell your people to head in that direction (toward some holes).

Dulcinea: Yes, it was the last place of that signal.

Sergeant Martinez: We... (static noise)

Colonel: Oh no, again, the same static noise.

Alexandrus: Oh oh, I think they're in trouble.

Dulcinea: Sir, this time believe me, get them back now.

Colonel Demello: Sergeant Martinez, get your people back...it's an order.

Sergeant Martinez: (static noise) Sir, I can't hear you very well.

Alexandrus: No this time, I'm going. We're not going to leave these soldiers stranded.

Amadeus: Me too.

Atlacatl: I'm in, let's go.

Colonel Demello: Where are you going?

Alexandrus: We're going to help them. Something is out there, I don't know what, but they need our help.

Our heroes left the base and with them were others soldiers too. They got another combat rafts, and were hearing gunfire. They were on their way to the rescue.

Alexandrus: They're firing. I hope we make it in time.

Amadeus: Look over there.

They saw Sergeant Martinez running to the shore, trying to leave the place. He was afraid of something.

Sergeant Martinez: Let's go, people, there's something weird.

He fainted to the ground, wounded.

Alexandrus: Amadeus, take him to the base.

He took off his mask.

Amadeus: Look at his feet, they're bleeding

There was too much blood on his ankles.

Dulcinea: Looks like he was bitten.

She took her mask off too.

Amadeus: Bitten, by whom?

Alexandrus: It's why we're here, to find out, so watch out, guys.

Dulcinea: You too. Alexandrus, something is wrong here.

The place was foggy, it was very difficult to walk, and there were screaming voices.

Alexandrus: Look at that tree, it seems that somebody or something was there.

There were something like bear's claw marks. They were big.

Dulcinea: It's a trace of a big animal…a bear? In this place?

Alexandrus: I don't know, but look, there's a blood trail. I'm following it.

Amadeus: Here too.

Atlacatl: Over here too. They disappear in the middle of the… shore. It looks like the sand swallowed them.

They heard the screaming voice of a soldier.

Alexandrus: Over there—hurry up, guys!

Now somebody else was screaming for help. They had no idea what was waiting in this new world.

Dulcinea: What's going on? I don't see the enemy.

The worst of all was coming. They didn't see the soldiers, no trace at all.

Alexandrus: Where are they? It's here, they just vanished.

All they're seeing is the wind blowing the palm trees.

Dulcinea: I don't like it.

Alexandrus: They disappeared. That was fast.

Amadeus: But how?

They found themselves in a quite different world from ours, and something from the ground was emerging. Two long arms jumped from the sand and grabbed Alexandrus's feet.

Alexandrus: What the heck?

Dulcinea: Watch out, Alexandrus.

Amadeus: They got me too.

Dulcinea: Here's my knife.

She threw her combat knife to Alexandrus, who caught the knife and immediately cut the two long arms.

Amadeus: Take this, you son of bad…

The other soldiers were under attack too.

Dulcinea: I'm coming.

Alexandrus: Here's your knife, I got mine.

He threw the knife in the air, and she caught it went to help them. They were just cutting those hideous long arms. So they all started to fight until the worst came up from the bottom of the ground. There were strange creatures jumping into the surface; they were very scary, and they looked like Roman warriors with swords and shields but without faces. They had long arms, metal masks on their faces, and a strange voice of communication. They were the "Topos," a savage tribe that lived underground and ate humans. They were so powerful.

Alexandrus: What are those creatures? Why haven't we been told about them?

Amadeus: I don't know, but let's finish them off.

The Topos were attacking our people furiously.

Amadeus: Take this!

He grabbed the big sword from one of them and killed him. Three others jumped off. The swords were very sharp and they stood watching; this was real sword fighting. They had to slash at their enemy's legs because they had no armor.

Dulcinea (watching the others): We have to strike their legs.

She jumped with both feet off the ground so her blow went under them. She got one and Alexandrus got the second one; the third one escaped under the ground.

Alexandrus: Guys, are you all right?

Dulcinea: We're fine.

They were covered by green blood and sand. At that moment Colonel Demello came with some help.

Colonel Demello: Are you okay? What happened?

Alexandrus: We're fine, but what we did is just the beginning. We better keep going; they know that we're here. We better hurry up and finish this quest.

Colonel Demello (looking at a fallen Topo): What's that? This mother f… is ugly.

Professor: Incredible. It's a kind of mutant from… I don't know.

Dulcinea: They're some kind of DNA experiment. This was true, they did it. We better leave now.

They left.

Colonel Demello: Okay, listen up. I want all of you to unpack everything. Sergeant, your mission is to try to communicate with the others. Campbell, make a perimeter; Jordan, take your men on the lead and give me a report on casualties.

Meanwhile Alexandrus and the others are talking.

Alexandrus: Hey, what was that?

Dulcinea: I'm not sure yet, but I think my parents mentioned some kind of experiment where their scientists tried to prove that they could live underground.

Alexandrus: I see.

Amadeus: Yes, but in this case, the experiment went wrong.

Atlacatl: I don't get it, the Zurdos made all this. They combined DNA from animals and humans, and God knows what else.

Dulcinea: Their goal was to achieve super strength on their bodies. They tried to use it in medicine but I guess something didn't work.

Alexandrus: The same thing on Earth—our scientists are making the same mistake, trying to be God.

Amadeus: And they said in the name of medicine and future, blah, blah, blah.

It was suddenly quiet. Colonel Demello and Captain J.J. approached our heroes.

Colonel Demello: Hey guys, any idea what we're dealing with?

Dulcinea: I thought that you knew, Colonel.

Captain J.J.: Come on, it's not the right moment, you better speak up.

Colonel Demello: Relax, Captain, leave them alone.

Alexandrus: What we think you probably won't like, but we're in trouble.

Colonel Demello: What do you mean? I know we're in trouble, son.

Amadeus: What Alexandrus is trying to say is that not only we are out of equipment, ammunition, soldiers; we are out of food and water.

Dulcinea: And that's not all, we're running out of time.

Alexandrus: In case you haven't noticed, guys, we're lost too,

Atlacatl: One more thing, guys, we don't know exactly where the enemy is.

Colonel Demello: So we're doomed, right?

Dulcinea: Kind of.

Alexandrus: Colonel, I think we should stick together. We have to regroup; we've traveled for so long and we'll need a plan.

Sergeant (holding a computer in his hand): Colonel, Colonel, I'm sorry to interrupt you, but I think you should see this. Colonel Demello: Go ahead, Sergeant.

Sergeant: I don't know what's happening here, but I haven't received any signal at all, our radio transmission doesn't work, and our computers are down. Our communication is useless.

Colonel Demello: What? Are you serious?

Sergeant: Yes sir, I think the surface of Europa or Jupiter has something to do with this.

Alexandrus: Both, Colonel. It's useless to use this equipment here.

Captain J.J.: Colonel, now what? We're sitting ducks, sir, we should do something.

Colonel Demello: I know, Captain, I know, I'm thinking.

Alexandrus: Relax, gentlemen, relax, we'll come up with a solution.

Colonel Demello: We're serious about this. I think we're going to have a meeting with our personnel.

Alexandrus: Colonel, why do you think we were brought here? Because we won a vacation?

Captain J.J.: Stupid assh…get to the point.

Alexandrus: Hey, hey, knock it off, Captain. I've had enough of you.

Dulcinea: Relax, Alexandrus, don't do this. Colonel (she turned to face him), we are the solution, don't you remember? We can communicate through telepathy; we're connected in some other way with our minds.

Colonel Demello: Kids, you never stop amazing me. How could I miss this?

Alexandrus: Okay, I think we should split in two groups or three. In group A, Colonel Demello and half of your people, sir, including me; and group B, Captain J.J. can take Amadeus and Dulcinea and the rest of our people. We're going to the frontline and you guys are going to attack from the back.

Amadeus: Aha! I mean how? We don't have our heavy artillery.

Colonel Demello: The artillery is us. Okay, everybody, listen up.

Colonel Demello started to organize the people in each group with Captain J.J.

Captain J.J.: Sir, with all due respect, I don't think it's a good idea.

Colonel Demello: Do you have one, Captain?

Captain J.J.: No sir, not yet.

Colonel Demello: Okay, go to work. We have to save whatever is left.

After a brief meeting, everybody was trying to save the few things useful from the landing crash. They were in big trouble, and the sunlight was fading away.

Colonel Demello: Okay, ladies and gentlemen, we're going to camp here and tomorrow we'll start again.

The night was so cold that it was really hard to start a fire. The orange and dark purple sky was so gorgeous that it was like they were in a dream. They could see Jupiter and two more moons (Calisto and Guminade), and the landscape was beautiful. The spot that they chose was completely safe...for the moment.

Colonel Demello: Okay, people, we're going to rest for a while; back on our planet it's night, but here I'm not sure if we're in day or night. I only know one thing for sure: it's going to be the hardest part of our mission. Whoever stands at the end of this mission the people from Earth will be glad, and our generations will see the future and they will dream again.

In the same spot of the group were our heroes.

Alexandrus: Okay, guys, we'll be in a tough mission from now on and we'll be depending on every one of us. Dulcinea and Amadeus, be careful. You know, guys, this is the only way we can survive.

Dulcinea: Don't worry, Alexandrus, I know this is our only chance, but I don't like the idea of going somewhere with Captain J.J.

Amadeus: I agree with you, Dulcinea, but Alexandrus is right. Besides, Alexandrus is the one Captain J.J. hates.

Alexandrus: Calm down, guys, I know Captain J.J. hates me, but I don't care. There's worse things to come. Just be careful with all the danger out there.

Atlacatl: Hey, what do you think is going on out there? I mean, do you think the Zurdos are waiting for us right now?

Alexandrus: I don't know but sooner or later we'll find out.

They spent that precious moment of resting talking about the mysteries and danger from Europa and the imminent conflict between the Zurdos and our troops. Professor Kishimoto was carrying a video camera, filming them. After a couple of hours of resting and not sleeping, the call was made for them to get going.

Colonel Demello: Okay, we know already that we're going to split into groups. Group A is under my command, and Captain J.J. is in charge of Group B, so let's begin the dance, people, and remember that we will stand until the end.

Everybody: Yeah, for Earth!

Colonel Demello (holding a map): Captain J.J., you take the north view and I'm going to take south; we'll meet here in this spot soon, and God bless you all.

Effectively, Group A was made up of Colonel Demello, Alexandrus, Atlacatl, the professor, and about sixty soldiers, and Group B had Captain J.J., Dulcinea, Amadeus, Lt. Dago, and about eighty soldiers. They were walking, keeping the good faith in God. Some of them looked nervous, and some of them looked sharp, like every situation of war was like a final chapter of their lives. They shared good moments with each other until they split.

In Group A, after a long journey:

Colonel Demello: Okay, Alexandrus, do you have any idea where we are going?

Alexandrus: Sir, I'm sure we're heading to the Zurdos, but I don't like the weather; there's a lot of rain. This could be bad for us. They can attack us anytime.

Colonel Demello: And do you think those creatures from the shore when we landed are still out there?

Alexandrus: I'm positive that they're waiting somewhere, just waiting for our most minimum mistake to attack.

Atlacatl came up.

Atlacatl: Sir, my men found something over that hill and I think you should take a look.

Colonel Demello: Well, let's go and be alert—we don't know what that is.

With some difficulty they headed right away to that place, and everybody was steady, alert, ready for the risk as they walked the difficult road in Europa.

Colonel Demello: What's that?

Atlacatl: Sir, it looks like a net, but I don't know what that thing is hanging.

Surrounded by fog, there was something like a spiderweb in the few trees that were standing in the entry. The cave was approximately 12 feet high, and it was really difficult to see through.

Alexandrus: Sir, this place could be a trap.

Colonel Demello: I know, son, but we need to find out what that is.

They approached.

Colonel Demello: Sergeant, take position of your men.

Sergeant: Yes sir.

They were moving with their bodies close to the ground and entered the cave.

Sergeant: I can't see well, there's too much fog.

He put on his night mask.

Colonel Demello: Just be careful.

Alexandrus: What's this? I think I heard something, bring a lamp.

Atlacatl: Take mine, it's much stronger.

Alexandrus: Thanks.

What they saw was indescribable. They found a pile of bones, and something else. The odor was very strong.

Alexandrus: Oh my God.

Atlacatl: They look like human bones

Alexandrus: Not only human bones, there's something else,

Atlacatl: Like what?

Alexandrus: I don't know, it could be animals, mutants, or even Zurdos. Let's get out of here.

They heard a noise getting closer.

Atlacatl: I'll take my men out, Alexandrus.

Alexandrus: Okay, I'll take the colonel out. Hurry up.

They started to run.

Alexandrus: Everybody out! There's something in here.

It was a very terrifying moment, and everybody heard a scream from a soldier.

Soldier: Arrgh, help!

Another one was screaming too.

Soldier #2: What the... Arrgh!

There were some shouts inside the cave.

Colonel: What's going on, Sergeant? Tell me what's happening!

Sergeant: I don't know, sir, there are some shots. It's really hard, I can barely see through the smoke.

Colonel Demello: Put on your masks and hold your fire, repeat, hold your fire.

But the screaming of soldiers one by one continued; they were shooting everywhere, and they shot some of the other soldiers.

Alexandrus: Atlacatl, are you there?

Atlacatl: Yeah, right behind you.

Alexandrus: Take your position, we're almost out.

Atlacatl: Yeah, I can see a light, let's go.

When they came out from the cave, they saw a puddle of blood and mutilated human parts on the ground.

Alexandrus: What this?

Atlacatl: Oh my!

Atlacatl called his men.

Atlacatl: Ramirez, Funez, Peraza, take the left right behind the rock. Lopez, Medina, point to that tree, and if something moves, just shoot.

Alexandrus: They killed the soldiers left in the entry, sir.

The colonel was coming.

Colonel Demello: Are these freaky mutants from the beach attacking us again?

Alexandrus: No, I don't think so, it's something up there.

Colonel Demello: What? Tell me. Oh my god, all the soldiers in the entry are dead.

Alexandrus: I can't see it, but something is up there.

And all of sudden they started to hear the sound of something blowing the air and lifting the dust from the ground; it was like a buzzing from a bee, but big.

Alexandrus: Did you hear that?

Colonel Demello: Yeah, what was that?

Alexandrus: Sshh!!

Colonel Demello (in a low voice): Please, can someone tell me what that was?

Atlacatl: Alexandrus, it's coming again, and this time it's getting closer.

Alexandrus: I know, just be alert, and Colonel, tell your men not to shoot.

Colonel Demello: Okay, hold your fire until my command.

Once again there was the buzzing sound, but this time it was taking one soldier one by one every minute, and they were firing back at it.

Alexandrus: Sir, tell your men not to shoot, we're wasting our ammunition.

Colonel Demello: I said hold your fire.

The silence was ruling at that time; nobody moved but Alexandrus. He had a big sword in his hand, and he was ready for anything. As he walked through the smog, he could see the soldiers' faces hidden in the bushes, stones lying on the ground. The buzzing was coming again. It was getting closer, coming right at him.

Alexandrus: Come on, I'm here…come on.

The buzzing was right behind him.

Alexandrus: Oh my!

He went cautiously down, and the buzzing missed Alexandrus's jump.

Alexandrus: The next move I'll catch it. It almost caught me.

When the buzzing got closer, he made an amazing move with his body; he turned around, held his sword in front of him, and all of sudden he spread out his legs into the ground and raised his arms into the sky.

Alexandrus: Take this, ahhh!

He cut it in half.

Atlacatl: You got it, Alexandrus, you got it.

Colonel Demello: Yeah!

Atlacatl: Are you okay?

Alexandrus: Yes, thanks.

Colonel Demello: I've never seen such movement like that in my life.

Atlacatl: What's that?

Alexandrus: It's like a big bee or mosquito, or both.

Atlacatl: It's like an experiment from the Zurdos.

Alexandrus: Not at all, I think this creature is from Mother Nature from Europa.

Colonel Demello: You're telling me that there are more of these creatures.

Alexandrus: It could be, but not in this area. I saw inside of the cave the nest is destroyed. I don't know who destroyed it, but it's gone.

Colonel Demello: Thank God it's over.

Alexandrus: It's not over yet, sir, it's just the beginning.

Colonel Demello: You're right, we're just starting. Sergeant, give me a report, how many casualties.

Sergeant: Yes sir. (He left.)

Alexandrus: This creature was just protecting its cave.

Atlacatl: From who?

Alexandrus: Something more dangerous, but don't ask me because I don't know.

Colonel Demello: We should leave immediately.

Alexandrus: I agree with you. Let's move—this is not a safe spot here.

Sergeant: Sir, we've counted fifteen dead and two missing.

Colonel Demello: I can't believe we're just getting down little by little and the real battle hasn't started yet.

Alexandrus: Well, that was a challenge before we left Earth. Every one of us knew the risk of this mission, so we must keep going and not let those soldiers die in vain.

Atlacatl: So what next?

Alexandrus: I'm not in charge; he is. (He pointed with his head at the colonel.)

Colonel Demello: Mmm, well, in this case, I think we should keep going, right, Alexandrus?

Alexandrus: Right, Colonel.

Then everybody left that freaky cave, but not before they held a small ceremony for all those who died.

Alexandrus: Let's go, there's more danger out there...waiting for us.

Atlacatl: Let's go. Okay soldados, vamonos (let's go).

Colonel Demello: Sergeant, let's leave now and be sure not to leave anything here.

Sergeant: Okay.

Meanwhile in Group B, Captain J.J. and the rest of his people were advancing by foot, but we have to remember that the special transportation for this quest was in the Desert Whale, and Europa's soil did not allow such privilege like that, so they were walking and walking until it was getting dark. They were also hungry; they had lost everything in the crash landing. They didn't know if it was a sunset or a sunrise like on Earth, but in that moment something intrigued our people.

Lt. Dago: Captain, did you hear that?

Captain J.J.: Hear what?

Lt. Dago: I'm not sure yet, but it's something.

Dulcinea: Yeah, I heard it too.

Amadeus: Me too..

Sergeant Roxford came up.

Sergeant Roxford: Sir, with your permission, we found some railroad tracks behind that hill,

Captain J.J.: What? A railroad?

Sergeant Roxford: Yes sir, you should take a look.

Captain J.J.: Let's go, Sergeant.

Quickly they went to see what Sergeant Roxford found. There was a long railroad track, they didn't see where it began and they didn't see where it ended.

Dulcinea: Amazing, it's like on our planet.

Captain J.J.: Yes, just take a look around; we don't have freaky monsters living underground.

Dulcinea: Not now, Captain, but if we continue doing experiments with DNA, someday we will end up like them.

Amadeus: Anyone can see that nobody has lived here for a long time. There's nothing here, not even animals. Something happened here, just watch out around...

Dulcinea: Right!

Amadeus: So, where is everybody? We haven't met any but those monsters back in the lake.

Dulcinea: Beach.

Amadeus: Or rivers, whatever, we should...

Lt. Dago: Shh! Did you hear it? It's that noise again.

Dulcinea: Yeah and it's coming right at us.

Amadeus: But where?

Lt. Dago: Over there, it's a light coming through the fog.

Dulcinea: What is that?

Captain J.J.: I don't know but Lieutenant Dago, get your men ready.

Lt. Dago: Yes, sir.

He turned back in position.

Dulcinea: I think it's a...train?

Amadeus: Yes, it could be a train.

Dulcinea: It's getting close, we should be in position.

The fog was making for poor visibility, but they were right, it was a train riding through the valley, heading right at them.

Captain J.J.: Hold your position until my command.

Lt. Dago: Yes sir.

He was giving orders to the rest of the soldiers.

Captain J.J.: (To Dulcinea and Amadeus) Any idea who this could be?

Dulcinea: I don't know, sir, but sooner or later we'll find out.

There's panic. The train was really close…and suddenly it stopped in front of them.

Amadeus: It stopped.

Dulcinea: Yeah and now what?

Nobody was getting off the train. It looked completely deserted but…

Amadeus: Nobody is in the train; I think we should go first.

Dulcinea: But why did it stop right in front of us?

Captain J.J.: Wait, we should make a move. Sergeant, make a perimeter around the train.

The soldiers started to surround the train.

Sergeant: Yes sir.

He was running with some others soldiers, and one of the car doors was sliding little by little.

Lt. Dago: In position, soldiers, a car is opening.

There was a creaky noise, and everybody got into combat positions, but the door stopped; it was halfway open.

Lt. Dago: Sir, and now what?

Captain J.J.: Send your people inside.

Amadeus: There's another door opening.

Lt. Dago: And another one.

Dulcinea: What's inside?

One of the soldiers was getting closer.

Soldier: Sir, there are two, three…oh my God, they're kids.

They were kids trying to get out of the car, but these kids were different. They were walking really slow, with their heads pointing to the ground, and the soldiers could not see their faces very well. Their hair was completely grey, and their clothes seemed really old and dirty. All of them were the same height. There were around five in the first car and ten in the other one; some of them were holding dolls and some kind of balls.

Dulcinea: They're kids, don't shoot.

Captain J.J.: Can someone talk to those kids? They need help.

Lt. Dago: Sir, we have Lt. Brown; our translator.

Captain J.J.: Bring him over.

Lt. Dago: It's she, sir. Lt. Brown, come over here.

Lt. Brown: Yes, sir.

Lt. Dago: You should try to talk to those kids.

The kids were walking in lines. They didn't approach our heroes, but they didn't lift their heads either. They were singing in an alien language, and for the soldiers it was a very creepy moment. They had no idea what was going on, and the kids seemed not to care; they kept walking.

Lt. Brown: It's all right, kids, what's your name?

The kids stopped.

Lt. Brown: It's all right, we came here to help you. (She began to talk in the alien language.)

Captain J.J.: (to Lt. Dago) What's that language?

Lt. Dago: It's the Europa language. We brought her here just for this kind of situation. She speaks ancient languages too.

Dulcinea: Wow, I didn't know she could speak the Europa language.

Amadeus: So, we're not the only ones now.

Lt. Dago: You speak it, too.

Dulcinea: Yes, I guess she learned it from our parents when they were in custody.

Lt. Brown was trying to talk to the kids, but it was in vain. Nobody knew what was happening until she put her arm on one of the children's shoulder.

Lt. Brown: It's okay.

The unexpected happened. The kid lifted his face at her and screamed really loud. He had very sharp teeth, and his eyes looked like they were bleeding. He grabbed Lt. Brown's arm and he bit her.

Lt. Brown: He bit me! I'm not sure what they are!

The rest of the kids lifted their heads, and they were the same. Their eyes were bleeding, their screams were very creepy, and their teeth were so scary—they looked like vampires mixed with zombies. They were so hungry, thirsty for blood, that they jumped into the soldiers.

Captain J.J.: Fire, kill those monsters!

The soldiers were firing at them, but they didn't die; they were still standing, they were still attacking and eating people, and they were fast.

Dulcinea: Do something, this is not right, they can't die!

Amadeus: Watch out, right behind you.

One kid was going to attack Dulcinea from behind.

Dulcinea: Geez.

She grabbed one of her special knives, which was like a sword, that her parents gave her before she joined the mission, and she cut the kid's head off.

Dulcinea: What are you? Take this.

Amadeus: That was a good move, but they're just children, I can't do it.

Dulcinea: Do it or you will die; they're not children anymore.

Amadeus: Here come some more.

There were three kids running at them, leaving dead soldiers in their wake.

Dulcinea: Amadeus, cut their heads off, it's the only chance.

Amadeus: But...

He had another special weapon that he brought from Earth; it was like a sword but different from Dulcinea's.

Lt. Dago: Sir, they're killing our men, we can't take them down.

The scenario was very bloody. There was no escape; they had to fight back.

Amadeus: Sir, cut their heads off. It's our only chance to survive.

Captain J.J.: It seems that...it's working

He grabbed one.

Lt. Dago: Cut their heads off.

And he grabbed his knife and was doing the same. It was the only way to stop these freaky kids. There were so many soldiers lying on the ground, some of them injured, some of them dead. The battle was almost in favor of our soldiers.

Dulcinea: We're almost done. That monster is the last one.

Captain J.J.: Leave it to me.

He grabbed his pistol and he shot all his ammunition right at the monster's face.

Lt. Dago: He was the last one.

Captain J.J.: What were they?

Dulcinea: I don't know, probably one of so many experiments, but like I said, I can't stand this. They were just children. The Zurdos have no mercy.

Captain J.J.: Lt. Dago, make a report on all casualties that took place

Lt. Dago: Yes, sir (He went right away to the wounded ones)

Captain J.J.: This is not what I thought, I was expecting that the Zurdos were something else, but kids…

Amadeus: Don't worry, you won't be disappointed. This is only one of their experiments.

Dulcinea: I don't get it.

Amadeus: What?

Dulcinea: Why haven't we found any trace of the rebels? I mean, we've just found freaky monsters and we're still losing soldiers; we still have to face them.

Amadeus: Probably we're in the wrong place at the wrong time.

Captain J.J.: Okay everybody, we should keep going.

Dulcinea: How about the dead ones? Are we going leave them here?

Captain J.J.: Of course, carrying them will slow us down.

Dulcinea: Yeah, but it's unfair.

Captain J.J.: I gave my orders, so keep moving.

Dulcinea: Yes, sir.

She couldn't believe how hard Captain J.J. was on them. They were leaving the dead ones. They were just carrying the wounded ones, who were crying from pain.

Dulcinea: It's funny that our weapons don't work here.

Amadeus: Yeah, it seems that our swords and knives are the only weapons that we need.

Dulcinea: I think so, too.

Amadeus: At least the soldiers down needed some kind of respect. We're losing men everywhere, and we haven't found the real threat yet.

Dulcinea: Yeah, that worries me too; by the time we get to the Zurdos we'll be outnumbered by them.

Amadeus: But we've got to finish it.

Dulcinea: You're right. I'm just wondering how Alexandrus is dealing with the rest of our people on the other side.

Amadeus: He will be fine. I've been having a hard time trying to communicate with him.

Dulcinea: Me too, it looks the Jupiter's gravity force is blocking us.

Amadeus: I hope they're fine.

Dulcinea: Me too.

Back at Group A... They found a new path, and the soldiers had walked a long way. They were really hungry and thirsty, but they had to managed to live. They were beginning to get tired when they noticed something.

Alexandrus: Hey, this road changes here.

Colonel Demello: There's no road, just volcanic rocks.

Alexandrus: Sir, we should be careful, I don't like the idea of crossing this way.

Colonel Demello: Me neither but we have to.

Atlacatl: There's no sign of anything.

Alexandrus: Yeah, that's why I don't like it.

Atlacatl: Any other choice?

Alexandrus: Well, this is the only way; we have to pass it until the other hill.

Atlacatl: I don't like that fog, it could cause a lot of difficulties. We might be sitting ducks over there. Remember that we're out of food and time.

Colonel Demello: Hey, I think you're right, time is priceless here. Let's go.

Alexandrus: God bless us all. (He was one of the first ones to walk.)

Colonel Demello: Sergeant, tell your men to be ready, I don't like this area.

Sergeant: Yes, Colonel.

The area was covered by fog, and walking through volcanic stones was very hard. Even with their masks they couldn't see well.

Alexandrus: Sir, we should stick together, I feel something very strange.

Colonel Demello: What is it?

Alexandrus: I'm not sure yet, but we'll find out soon.

Atlacatl: Just be alert.

Suddenly the fog was covering our heroes, and they barely could see each other. There was silence and loneliness until an unexpected noise broke the silence.

Soldier: Ahhhh!

Colonel Demello: What was that?

Alexandrus: It's a scream from somebody.

Then again and again, there were some shots.

Colonel Demello: It's a trap.

Alexandrus: Just stick together, don't run.

Colonel Demello: Everybody just stick together, and don't open fire.

Alexandrus: Can you hear me?

Atlacatl: Yes Alexandrus, I can.

Alexandrus: Remember when we were kids playing with wood swords?

Atlacatl: Yeah, but I don't get it.

Alexandrus: We might play again but this time for real. Tell your men to grab their swords or knives. I think we're not going to need guns in this battle.

Atlacatl: Hey, muchachos, agarren sus espadas y cuchillos y esten listos. (Hey, guys, grab your swords and knives and be alert.)

Colonel Demello: I said don't run!

The fire, the moaning, and the screaming were the result of that ferocious attack throughout the volcanic stones.

Atlacatl: Alexandrus, what are we fighting?

Alexandrus: I'm not certain yet, but I sense the enemy is not using guns.

Colonel Demello: What? So with what are they attacking us?

Alexandrus: I don't know yet, but they're not the monsters from the shore.

In that moment they saw a soldier get killed; he was stabbed by a metal thing through his stomach, and then it was pulled out right away.

Atlacatl: Did you see that?

Alexandrus: Yeah, just watch out.

And then again that metal thing jumped from somewhere,

penetrated the body of another soldier, and then it vanished.

Colonel Demello: Wow, can anybody tell me what that was?

Alexandrus: Shhh, sir. They're here, just be quiet.

The fog was disappearing, the scenery was getting a little bit clearer, and finally the Zurdos showed up, a battalion of approximately 30. Their weapons were strange swords shapes, and the Zurdos looked like humans in uniforms, mostly gray and black. Some of them were carrying their own weapons, and some of them looked like their swords were part of their bodies—those Zurdos looked very intimidating. Even though they were outnumbered by our heroes, they didn't care. Behind those warriors was another entity who seemed to be the leader, but in the fog it was impossible to recognize him at first sight.

Colonel Demello: What are those warriors?

Alexandrus: May I introduce you? They are the Zurdos.

Colonel Demello: Okay, what are we waiting for? Let's rock and roll, let's show them what we're made of.

Alexandrus: Atlacatl, take your soldiers and let's go for our planet.

Atlacatl: Vamos, yeah!

That was the beginning of the first battle against the Zurdos. Some of them were using guns, swords, knives, everything. Alexandrus, Atlacatl, and his men were using swords. The Zurdos were so fast and very smart; the soldiers were trying to shoot at nothing. They were an easy prey, and some of them were very lucky, but Alexandrus, the colonel, and Atlacatl were more successful in the battle. When they were kids, Alexandrus's father taught Alexandrus and Atlacatl how to use swords, and that was very helpful to our heroes.

Alexandrus: Take this! Colonel, behind you.

Their swords were clashing in the air.

Colonel Demello: Thank you, son. Take this, <u>cowards</u>

Everything looked horrible. The Zurdos were a well-trained army too, and they were very barbaric; they were not leaving anybody alive.

Alexandrus: Atlacatl, take your men back; the rest of the soldiers need our help, and I've got to get going.

Atlacatl: Are you sure you don't need help?

Alexandrus: I'm sure, go… They might need our help.

Colonel Demello: I'm coming with you, son.

Alexandrus: Sir, it's very dangerous, I've got to get to their leader.

Colonel Demello: Well, I'm coming.

They were fighting to get through to the leader but they were having trouble. They got closer and closer, but there was still one more fighter to get through.

Colonel Demello: You go ahead, son, I'll take him.

Alexandrus: Okay, be careful, sir.

Colonel Demello: I will.

Colonel Demello was fighting in hand-to-hand combat with this Zurdo, and Alexandrus went after their leader.

Alexandrus (to himself): Where are you? Come on, I know you are here somewhere.

The fog was still heavy, and it took him longer than he expected. He heard a sound that something was coming right at him. It was a blade again, but this time it missed, thanks to a wonderful move from Alexandrus.

Alexandrus: Oh my, it almost got me.

Once again that blade attacked from nowhere.

Alexandrus: Okay, you missed again, the next time I will guess it.

He continued to focus on his status as the fog increased. He slowly grabbed his knife without pulling out, and he heard that noise coming again.

Alexandrus: I caught you.

He threw his knife towards the noise, and he heard a scream. Alexandrus finally caught the blade. The fog was vanishing until they were alone, face to face, mano a mano. The Zurdos leader was grabbing one of his arms. He was Alacr.

The scenario was just for them. Finally Alexandrus was in front of Alacr, and they were immobile, studying each other.

Alacr: (in an alien language): What are you people? You're invading our soil, and I'm here to stop you. You are going to die.

Alexandrus: (alien language) We're from planet Earth and we're going to make all of you pay for what you did to our planet.

Alacr : Ha ha ha..

He threw his blade again, but this time Alexandrus saw that the blade was coming from one of his feet.

Alexandrus: What are you? Some kind of scorpion?

Alacr made another counterattack from the other foot, and he launched another blade. Each time he threw a blade, another appeared from his feet. They fought a brutal fight. Alexandrus's moves were so elastic, fast, and accurate, and in one of those moves, he was wounded in one of his arms and he fell down. He also was bleeding a little bit from one of his ankles from a Topo's attack.

Alacr: Ha,ha,ha you die.

Alexandrus made a quick move and avoided the contact from one of Alacr's blades; then he found a sword lying on the ground from one of Alacr's guards. He grabbed it and cut off one of the blades.

Alacr got angry. He grabbed his sword and they fought, and then again the fog was coming in. Alexandrus was at a disadvantage; he didn't know the place and again from nowhere, the blade was launched. Alexandrus avoided and was steady, just waiting for the next attack. He had his sword in his hand and he was looking around, bleeding from one arm. He was tired but he knew it wasn't time to quit. He heard from far away the gunfire from his friends, and he knew they were having their own battle. He attacked again, but this time Alacr used his sword. Alexandrus realized that Alacr was bleeding from his shoulder, and this gave Alexandrus courage to keep fighting.

Colonel Demello was having trouble taking down his rival. It seemed that he was overpowered by this Zurdo, but then the inexplicable happened. When the Zurdo had a chance to kill Colonel Demello with his sword—Colonel Demello was lying on the ground, tired and accepting his fate—he said something in his language. Of course Colonel Demello didn't understand a word, but the Zurdo continued talking, and Colonel Demello was surprised.

Colonel Demello: What? You're not going to kill me?

The Zurdo said something else.

Colonel Demello: I wish I could understand, but I don't speak your language. (Softly) I wish it could be like in our movies, and every time we visit another world or the aliens come to ours, they speak English.

Surprisingly he was interrupted by the Zurdo, who dropped his sword and gave his hand to Colonel Demello, to help him get up.

Colonel Demello: What are you going to do to me?

The Zurdo helped him to stand up, and then he tried to communicate, making signs with his hands and talking in alien language.

Colonel Demello: Okay, I don't understand a word, but thank you for sparing my life. I'll take you to Alexandrus; he will know what you are saying, because I don't understand any word.

Meanwhile Alexandrus was in trouble. He was getting tired and wasn't getting any breaks until at one point, he looked as if he was going to lose the battle. Alacr was a good warrior, and he didn't know that it was going to take a long time to defeat him, but again Alexandrus was really good too. He almost got caught with the enemy's blade, but he fought back with excellent technical moves. His sword touched part of Alacr's stomach, and once again Alexandrus made sure that it was his chance to take. His sword clashed with Alacr's sword and both dropped off.

Alexandrus was caught with a hard hug, and he saw that Alacr was going to use his blade. He made his last effort to get out, and hit Alacr's head. It was at that moment that his own blade crushed him, and he fell down. Alexandrus had won the difficult fight. Alexandrus was exhausted, bleeding and tired, and then Colonel Demello showed up with his new ally.

Alexandrus: Colonel, you have one of them!

Colonel Demello: Relax, son, he's my friend. He insisted that I spend Father's Day in his house. He's cool. I like this guy.

Alexandrus and this Zurdo were looking at each other. Colonel Demello was still talking about his new friend.

Zurdo: (alien language) You must be the son of Curru.

Alexandrus: (alien language) How do you know my father, and who are you?

Zurdo: My name is Miche. I'm the second commander from this region—well, I guess I'm first since Alacr is down. We were sent to kill all of you.

Colonel Demello: What's going on? I don't understand anything; what are you saying?

Alexandrus: And then what happened? Why have you betrayed your people?

Miche: I haven't betrayed my people. My people, my real family, live like slaves, in chaos, without hope. After the Zurdos took control of this place, they changed everything. They are merciless. They've killed everybody in their way and now they're planning on an invasion on your planet.

Alexandrus: Not if we act first.

Miche: On the first attack, something went wrong; they were expecting to kill the people but not the planet, but like I said, something went wrong.

Alexandrus: Not at all, they killed a lot of innocent people, and that's why we're here trying to stop these Zurdos.

Miche: It's going to be hard; many cultures have tried before without any success.

Alexandrus: What do you mean many cultures?

Miche: Many civilizations, like the red planet.

Alexandrus: You mean Mars?

Miche: Well, we call it Marte, the red land.

Alexandrus: I know that word; it means "War God."

Miche: Well, a long time ago, they were a great civilization, maybe one of the most powerful in this galaxy. Our race originated from Jupite.

Alexandrus: Jupite? We call

Miche: They were in danger; Jupite was hit by a big storm of meteorites, and by that time their scientists had made this place habitable, but they didn't have a choice. It was very small, and relocating all of them created chaos. There were two tribes, always fighting each other, and they wanted to have control of that place, but after the imminent danger of that planet, they united forces to get these people out. Not all of them made it alive. The Zurdos chose their people over us before they asked the Atlantis for help.

Alexandrus: Wait, what did you say? ATLANTIS!

Miche: Yes, Atlantis was a great civilization a long time ago. They were living in Marte, they were explorers, and they had an advanced technology.

Alexandrus: You tell me. I know who they were; they lived on our planet a long time ago.

Miche: The Atlantians denied their help, said that they couldn't deal with nature, and then suddenly they were going to help, I don't know what made them change their minds, but it was too late. By that time Jupite was being hit by a shower of meteorites, and only eight percent of their population made it alive, so after that period the Zurdos began their revenge on the Atlantians. When they arrived here, they found a powerful mineral capable of many things, and they created a powerful arsenal.

Alexandrus: Yeah, the Super Eurotanius. And they were ready to invade Atlantis.

Miche: The Atlantians never thought they would be attacked, and they were wiped out by surprise. The Zurdos wanted to take control of Marte but never thought with the extermination of Atlantis they were going to destroy the atmosphere too.

Alexandrus: Aha, now it makes sense that there's no life on Mars, but there was a long time ago.

Miche: So when the Zurdos realized that they destroyed the planet Marte...

Alexandrus: Let me guess, they found out that the next step was to take the Earth down.

Miche: Aha, they didn't want to destroy your planet either, they just wanted to kill your people and then to conquer it.

Alexandrus: Why are you telling me this?

Miche: Because my ancestors were the slaves of the Zurdos, and for that reason we sent Ingui, Une, and Curru to prevent that holocaust again and get some help from you guys. We want to live free again, like we did before.

Alexandrus: Well, I'm shocked. My name is Alexandrus, and I'm Curru's son. My father told me that one day it was going to happen, to defend our planet, but he never told me the whole story about the origins of the Zurdos.

Miche: Well, it's simple—they never knew it.

Alexandrus: I think you joined to help us find the Zurdos.

Miche: I will lead you to them. It's going to be hard; they have a lot of merciless warriors, loyal to King Atonal, and there's going to be many obstacles to get to the city,

Alexandrus: Don't worry, we're prepared for that.

Miche: I don't see you have a lot of warriors to contend with them.

Alexandrus: Well, we'll find a way to destroy those Zurdos.

Alexandrus turned toward the colonel, who was sitting down.

Alexandrus: Sir, we better keep going. I'll let you know all about this on the way.

Colonel Demello: Okay, son, tell me all about the conversation. Sergeant, let's go, let's get out of here.

They left, but not before holding a small ceremony for all those who died in the battlefield. Meanwhile, on the other road, Dulcinea and Amadeus, after a long journey of walking, were exhausted and thirsty.

Dulcinea: Hey, Amadeus. I can't believe it.

Amadeus: What?

He was eating a small portion of a candy bar.

Dulcinea: The Zurdos, just look what happened today with all those children.

Amadeus: Yeah, it's a shame, they're trying to stop us at any price. They don't care about using kids.

Dulcinea: They were just an experiment.

Lt. Dago came up from behind them.

Lt. Dago: Captain J.J. is ordering us to take a break. We've been through a lot today.

Amadeus: It was about time. Thanks.

Lt. Dago left.

Dulcinea (looking around): I don't like this.

Amadeus: I don't like what I'm hearing but you're right.

Dulcinea: Probably it's just my imagination.

Our heroes are nervous and spending the cold night was the worst. It had been a couple of hours since they chose that spot to rest. It was really cold and windy, and our heroes had made a small perimeter around them. They had a fire going, and some of them were sleeping in sleeping bags. Some of them were taking turns watching, and some of them were simply praying for their lives. Psychologically they were very devastated by all they had been through.

Soldier #1: Hey, it's my turn, go to sleep.

Soldier #3: Thank you, I'm sleeping.

Soldier #3 left. Shortly after, there was silence, and the only noise that they could hear was the wind from the trees' leaves.

Soldier #3 grabbed a cigarette and started to smoke. He looked around the perimeter where the soldiers were sleeping and turned to the other side, where there was nothing but bushes. Then he looked up in the sky and was amazed by the beauty of such a place. Suddenly he turned his head back to the bushes, and what he saw was scary.

Soldier #3: Oh my God!

He saw a bunch of dark shadows coming at them. He couldn't see what they were. It was really dark so he grabbed his flashlight and what he saw was the unexpected. He was in shock and shaking from fear. He saw all the dead soldiers, the same ones that died in the attack from the train.

Soldier #3: Hey, Mike, is that you?

He couldn't see their faces very well, but he saw the tag of one of the soldiers.

Soldier #3: Don't come any closer.

Then he saw that the soldier's eyes were bleeding just like the kids on the train.

Soldier #3: It can't be, no.

He opened fire at those dark shapes, and one of them went for him, killing him instantly with its teeth.

Dulcinea: What was that?

She had been sleeping in a small tent with three other women.

Woman #1: I don't know, we're under attack.

Dulcinea: Oh my God.

She stood up immediately and grabbed her weapons and she left.

Outside of the perimeter:

Dulcinea: What's going on, Lt. Dago?

He was opening fire with his gun.

Lt. Dago: Look for yourself. We're under attack by these vampires or zombies or whatever they are.

Dulcinea: They're the dead soldiers we just buried earlier today.

Lt. Dago: I know. Open fire, Dulcinea. They're coming.

Amadeus: Oh, not these zombies again. We have to finish them this time.

He opened fire.

Dulcinea (also opening fire): We have to stick together; nobody leaves the perimeter. You might get lost and be easy prey.

It was really dark and our heroes could barely see each other. They were launching lights up to the sky to get at least a good chance to shoot, and Dulcinea was right—the only chance that they could survive that attack was by sticking together inside of the perimeter.

Dulcinea: Okay, listen up, everybody. Just fire at whatever you see, and don't waste your bullets. Remember, we're short of ammunition. Amadeus, we can finish this. Take your sword.

Amadeus: Okay.

Dulcinea: Lt. Dago, point that light over those bushes. We're coming and don't shoot.

Lt. Dago did what Dulcinea told him.

Dulcinea: Thank you. Let's go, Amadeus.

Amadeus: Let's go, don't shoot.

They both jumped into the spot that Dulcinea saw was the worst. And that was it, our heroes attacked again. They were fighting with their swords, and there was a new kind of monsters—vampires, coming from everywhere.

Lt. Dago: Wait, don't shoot.

Captain J.J.: I can't believe those jerks are trying to be heroes.

Lt. Dago: Sir, with all due respect, I think we should be proud to be on their side.

Captain J.J.: Lt. Dago; don't make me feel bad, okay.

Lt. Dago: Sorry, Captain, but I'm going too.

Lt. Dago joined the battle.

Lt. Dago: Take this, bastards!

Lt. Dago opened fire. Surprisingly there were more soldiers jumping from the bushes, and the threat was dire. Zombies and vampires were also taking soldiers down. Then it started to rain, and the wounded soldiers began to act differently. They turned into vampires—zombies too. They were inside of the perimeter, and there was no escape. One of the nurses was bitten.

Nurse # 1: Oh my!

Guard #1: Here, medics, we have monsters inside of the perimeter.

Sadly, another battle began inside of the perimeter.

Guard #5 (to Captain J.J.): Sir, they're attacking us.

Captain J.J.: Who?

Guard #5: The wounded ones from the first encounter.

Captain J.J.: Get your men in position and let's go inside.

Guard #5: Yes sir.

They went immediately to where the wounded were recovering, and when they arrived, they saw chaos. Some soldiers were using their guns, some of them were lying on the ground, and there was no chance to leave; they had to kill them or be killed. The battle was inside and outside; it was intense and bloody.

Amadeus: Take this, I got nine.

Dulcinea: I lost count.

Lt. Dago: I got ten.

Amadeus: We're almost done.

Dulcinea: We better finish quick. I've heard that in the medic room, they might need our help.

Amadeus: Okay, and take this, yeah.

Lt. Dago: That was the last one.

Dulcinea: Lt. Dago, take the wounded ones apart, seize them.

Lt. Dago: And the dead ones?

Amadeus: Burn them to hell, quickly.

Lt. Dago: What?

Dulcinea: He's right, you don't want them back.

Lt. Dago: Of course not, we'll do that.

Dulcinea and Amadeus left the scene.

Captain J.J.: Soldiers, to your right, open fire.

They were having a hard time. It seemed that the bullets were useless.

Amadeus: We're on our way.

In the medic room:

Dulcinea: Don't shoot, take your weapons down.

Amadeus and Dulcinea were using the only weapon useful, the swords.

Amadeus: Sir, take your men out and take the wounded ones apart.

Captain J.J.: Who the heck are you to give orders, little girl(to Amadeus).

Captain J.J. was really upset. He felt ashamed by those two young civilians, and he felt more pain, due to the fact that they were gaining admiration and respect from the rest of the soldiers.

Dulcinea: Sir, with all due respect (she was taking one vampire-zombie down), it's not the time for this. Please take the wounded ones apart.

Amadeus: Come on, sir, we have no time.

Captain J.J. ignored them; he started to throw a grenade where a monster was eating.

Dulcinea: Not there, there's flammable stuff there. Please, Captain.

Ignoring Dulcinea's pleas, Captain J.J. threw the grenade, which exploded in a big blast, and then the fire started. It also took down some soldiers on the other side. Captain J.J. left the site.

Dulcinea: Oh no, there are soldiers down.

Amadeus: I'll take the last two and go. Take whatever you can save.

Dulcinea went right away to the site and tried to save a soldier.

Soldier: Right here! Help!

Dulcinea: Come on, get up, you're not going to die here, okay? Get up.

She was trying to get him to safety but others were asking for help.

Dulcinea: Hold on, I'll be there.

She had to hurry because the fire was destroying the small room. She pulled one soldier up, and they were walking, but she couldn't help the other one.

Dulcinea: Oh my, I'll be back.

Amadeus: Don't worry. I'll take him out.

The fire spread quickly, and our heroes saved those soldiers from the fire, but they couldn't save the food and ammo. Saving their lives was most important, and when they were out, the rest of the battalion was cheering them on.

Platoon: Yeah, Dulcinea and Amadeus, yeah.

Dulcinea: Oh thank you, guys, but any of you could have done this too.

Amadeus: Thanks, thank you.

Captain J.J. was really jealous; he couldn't handle that Dulcinea and Amadeus were gaining everybody's respect.

Dulcinea (to Captain J.J.): Sir, I think we should leave.

Captain J.J.: What do you think, that I don't care? I'm in command here.

Dulcinea: I know, sir, but I'm just suggesting that...

Captain: I think you don't understand English. I'm the one who gives the orders.

Dulcinea: Yes sir.

Lt. Dago (approaching): Sir, we've taken the wounded ones and the dead ones too.

Captain J.J.: Burn them all.

Lt. Dago: But, sir, they're alive, we can't do that.

Captain J.J.: Do as I said, I don't want to say it again.

Lt. Dago didn't have a choice but to follow orders.

Lt. Dago: Yes sir, and we're out of bullets too; I think one or two rounds at most.

Captain J.J.: Tell our soldiers not to waste the bullets, just use what's necessary.

Lt. Dago: Yes sir.

And he left. Dulcinea and Amadeus went after Lt. Dago.

Dulcinea: Lt. Dago, what's going on? I think you should tell everybody to get knives or swords; we might use them.

Lt. Dago: Yeah, this is our only chance to survive.

Amadeus: Bullets in this place sometimes are useless, but knives are helpful.

Lt. Dago: Especially in fights between man to man.

Amadeus: Or man to whatever those things are out there.

Dulcinea: So how many casualties?

Lt. Dago: Well, we're not lucky this time. We just have three wounded from the fire, plus three who have been bitten and twenty down.

Dulcinea: I think we should find something for those injured ones.

Amadeus: Like what?

Dulcinea: I don't know, but I remember that my father told me once that in this place they have a plant that has miracle medicine; it heals anything.

Amadeus: Any idea what it looks like?

Dulcinea: Nope.

Amadeus: Well, there's no chance to find that miracle plant.

Dulcinea: Let's keep our faith.

Lt. Dago: Okay, I've got to go. I've got to burn them all.

Amadeus: What? They're still alive.

Dulcinea: We didn't come here to kill our troops, we're not killers.

Lt. Dago: Believe me, it's hard for me too, but Captain J.J. is pissed off because he thinks that they can come back for us soon, and we can't carry them the whole time.

Dulcinea: You're right, but we should do something soon.

Amadeus: Very soon, look to your right. There's a big storm heading towards us.

Dulcinea: We better find them a place to hide.

Amadeus: Yeah, but we have to lock them up.

Suddenly they heard some shots.

Lt Dago: What's happening?

They left right away.

Dulcinea: Oh my God, the wounded ones who were bitten are dead.

Lt. Dago: There's nobody here, they've killed themselves.

They wounded knew the risk against the others and took the option of killing themselves with their guns. Our heroes were speechless. The others were coming too, and some of them were crying. What they did was so brave—they wanted to keep doing their mission until the end. It was a sad moment and they couldn't believe it. Suddenly...

Dulcinea: Oh my God, there's a light.

She pointed at the top of the hill.

Lt. Dago: Not just a light, there are more over there.

Amadeus: It seems that it's a small village.

Lt. Dago: What do you think?

Captain J.J. arrived.

Captain J.J.: Hey, people, do you sense anything over there?

Amadeus: Like what, sir?

Captain J.J.: Anything. It's why you were brought here.

Dulcinea: You mean Alexandrus and the rest of them?

Captain J.J.: Yeah, it could be them.

In that moment it started to rain; it was getting to be almost daylight.

Dulcinea: I don't sense anybody.

Amadeus: Me neither but I suggest that we should go. This rain is cold.

Dulcinea: Yeah, with your permission, of course, sir.

Captain J.J.: Lt. Dago, let's get out of here; get your men in position and let's go to that village.

Lt. Dago: Immediately, sir. (He left.)

With some difficulty our heroes were heading to the village. It was raining, it was cold, and they were looking for some shelter. They didn't know the danger that was waiting for them.

Dulcinea: What do you think, Amadeus? Are there going to be people living in that village?

Amadeus: I don't know, but there might be. There's some light and I hope they have food.

Dulcinea: I hope they have some food, especially since we lost everything in the fire.

Amadeus: And some blankets, because ours are not enough.

Meanwhile Alexandrus, Colonel Demello, Atlacatl, and the rest of our heroes were getting ready for a new journey.

Alexandrus: Well, gentlemen, it's time to get up and go.

Atlacatl: Wow, what a long night. I thought that we were going to be in the dark the whole mission.

Professor Kishimoto: Well, gentlemen, we should blame Europa. It's the rotation around Jupiter. It takes 35 hours to complete its movement, so that means we're going to have long days and long nights from now on.

Alexandrus: Terrific.

Colonel Demello (approaching): Good morning, everybody. Get ready, we've got to get moving.

Alexandrus: Okay sir.

They packed everything that they had. It was raining.

Alexandrus (alien language): Hey what's up, Miche? Are we getting any closer to Atonal's city?

Miche: (alien language) Don't worry, we'll get there soon.

Suddenly they saw in the sky some big animals floating in the air that looked like whales. They were making a weird noise; it looked like they were calling each other, but they were harmless. It seemed like a flock of seagulls heading somewhere else, and their noise was like a soft music to their ears.

Alexandrus: Whoa, what a beautiful melody to our ears. It's amazing.

Miche: They are the Buchis. They're heading to the north for food. This winter is going to be very hard, not only for them but for us too.

Dulcinea and Amadeus arrived at the village.

Dulcinea: You know what?

Amadeus: Don't tell me, you don't like it, right?

Dulcinea: No, it's the lights, they're still on.

The village was completely empty, and it was raining. The shacks, or small houses, were empty, some of them with their doors opened. The light wasn't electricity at all; it was from "Kandils," a kind of a bottle filled with gas, and in the top was a piece of cloth with flame.

Amadeus: Their food still there.

Lt. Dago (from behind): It's empty.

Captain J.J.: Lt. Dago, I want a preliminary report; I want to know why it's empty.

Lt. Dago: Yes, sir.

Dulcinea: It's strange, where are these people? They had their food on the table, they were here, and I don't know, it seems that they just disappeared.

Amadeus: Don't you think it might be a trap?

Dulcinea: It might be, but still, it's strange. Hey, what's that?

Amadeus: Over there!

Dulcinea (pointing at one of the shacks): I think I saw something.

Amadeus: Like what?

Dulcinea: I'm not sure yet, but I'm going.

Amadeus: Should we tell the rest?

Dulcinea: No, let's go.

They went outside a small house.

Dulcinea: It was here.

Amadeus: But what did you see?

Dulcinea: It was something moving right here.

Amadeus: It could be something, an animal, one of our soldiers, or even the wind.

Dulcinea: No, it was something else.

They heard a noise inside of the house.

Dulcinea: Let's get inside.

They go inside.

Dulcinea: Be ready.

Amadeus: Always.

There was nothing inside of the house, until a door slammed against the wall.

Dulcinea: There!

They saw a little girl running, trying to get away from our heroes.

Amadeus: It's a little girl.

Dulcinea: She's scared.

Amadeus: Be careful, it could be a zombie-vampire like the ones we just fought.

Dulcinea: No, I don't think so, this time is different.

They follow her inside of the shack.

Dulcinea: (alien language) It's okay, we're here to help you, don't be scared. What happened?

They got to the girl, who was scared, almost crying, and she was around five years old.

Dulcinea: We're here to help you—what's your name?

Amadeus: (alien language) We're from another planet, we're here to help you. Where are your parents?

Simbra: (alien language) My name is Simbra.

Dulcinea: Okay, Simbra, my name is Dulcinea. Where are your parents?

Simbra: They're gone.

Dulcinea: Where? Are they coming back?

Simbra: No.

Dulcinea: Where are they, can you show us?

Simbra: I don't know.

Amadeus: Where are the rest of your people?

Simbra: They're also gone.

Dulcinea: Why are you here?

Simbra: I hid. I was so scared.

Captain J.J.: (arriving with three soldiers) What's going on?

Dulcinea: Nothing, sir, we just found this little girl here.

Captain J.J.: Take her prisoner, she could be a monster.

Dulcinea: With all due respect, it's a little girl.

Captain J.J.: I don't care, I've given my orders.

Amadeus: Sir, it's not a good idea.

Captain J.J.: Shut up, I'm in command, remember?

Amadeus: Yes sir.

Simbra: What's happening, Dulcinea?

Dulcinea: It's okay, Simbra, you're going to be okay.

One of the three soldiers grabbed her and they start to leave.

Simbra: Where am I going?

Dulcinea: It's okay, I'm going with you.

Captain J.J.: What the heck are you talking about?

Dulcinea: Nothing, sir, nothing.

They left the house, and outside something was expecting them.

Captain J.J.: Where are you soldiers?

Captain J.J. had been escorted by six soldiers—three inside of the shack, and three were outside, waiting for them.

Captain J.J.: I said, where are you soldiers?

Dulcinea and Amadeus knew right away, something was wrong.

Dulcinea: (telepathically) Amadeus, get ready. Something is not right.

Amadeus: (telepathically) I know, I don't like it.

Captain J.J.: Don't start with me, soldier.

He turned his face to one of his soldiers.

Soldier: (nervously) They're gone, sir.

Captain J.J. turned to the other soldier.

Captain J.J.: You too. Leave the girl with Dulcinea.

The soldiers were looking for strategic points around the area.

Soldier: Yes sir.

He left. The rest of them stayed outside the shack. This small house was the last one from that village, but they had to come back where Lt. Dago was waiting for the signal.

Dulcinea: Sir, I think we should get moving.

They heard gunshots.

Captain J.J.: What was that?

Amadeus: Gunshots.

Captain J.J.: I know that. Where?

Amadeus: It seemed it's happening in the first house, sir.

Then they heard a scream and there were some shots.

Captain J.J.: And that?

Amadeus: That one is here, sir.

Captain J.J.: Get ready.

Amadeus: Like I said before… Always.

They were the first ones, walking slowly, slowly, and Dulcinea and Simbra were behind them. The rain still was hitting them hard.

Dulcinea: Sir, we should leave immediately, we don't know what's out there.

Captain J.J. didn't bother to answer.

Dulcinea: It's okay, Simbra, we're going to get you out.

They found traces of blood in the puddle.

Captain J.J.: What's happening here?

Amadeus: I don't know, sir, but we'll know soon.

Simbra: Be careful, they're really bad.

Dulcinea: Who? Simbra, who are they? Come on, Simbra, are there more people around here?

Amadeus, Captain J.J., the soldiers, Dulcinea, and Simbra were walking in the street very carefully. They didn't have any idea what was waiting further on the road.

Simbra: We should leave quickly. They're here, and they're going to take all of us, like my parents and my friends.

Amadeus: Hold on! Stop.

Captain J.J.: Why are we stopping?

Amadeus: Shhhh, we're surrounded.

Captain J.J.: By who? I don't see anybody.

Amadeus: Hey, Dulcinea, do you sense that we're not alone?

Dulcinea: Yeah, I feel it.

Simbra : It's too late, they caught us.

They were making a kind of circle; Simbra was in the center and our heroes had their backs to her, protecting her.

Dulcinea: (alien language) Stay back, Simbra, we're going to protect you.

Simbra (alien language): Nobody can protect us from them.

Captain J.J.: Where are they?

He took some pills for getting him high.

Dulcinea: They're here, sir, grab your weapon or whatever and be careful.

Amadeus: But where are they?

They were stuck in the middle of the shacks, and they were looking around but they didn't see anybody. They looked at the doors, roofs, even the ground, and nothing until there was a surprise.

Dulcinea: Watch out. There!

From nowhere a strange black shadow was spinning in the air and got close to Captain J.J.

Amadeus: Sir, be careful!

He pushed Captain J.J. onto the ground, and he put his sword near the captain's head.

Amadeus: Are you okay?

Captain J.J. was speechless. Amadeus's sword was standing between him and a big sword from this new black shadow, and it attacked again. Amadeus jumped into a new battleground.

Simbra (alien language): Please don't hurt us.

Once again from nowhere came another black shadow, spinning in the air.

Dulcinea (alien language): Stay there, Simbra.

She was fighting too. Our heroes were fantastic, amazing, and their skills were almost supernatural. They had known how to fight since they were children. And then the unimaginable happened.

Dulcinea: Take this.

She put her sword in the black shadow, which exploded into water.

Amadeus: My turn. I got ya!

His sword penetrated his rival's stomach, and again it exploded into water.

Dulcinea: What's going on?

Amadeus: I don't know.

Then, after a little pause, the black shadow appeared somewhere else, like nothing had happened.

Dulcinea: You gotta be kidding, you again.

And this time there were more, around eight.

Amadeus: Who invited those guys?

By that time Captain J.J. had opened fire at them with his gun.

Captain J.J.: What are those things?

Dulcinea: I don't know, sir, but I think the rain has something to do with them.

Amadeus: They're like water.

Simbra: (alien language) Dulcinea, they're the Water Knights.

Dulcinea (alien language): How do we kill them?

Amadeus: How do you kill water? You're kidding.

They're fighting again.

Dulcinea: I don't know, you're asking the wrong person.

Captain J.J.: I've taken down around three, and they keep coming again and again.

Dulcinea: There's got to be a way; we have to stop them and fast.

Amadeus: And we need backup, sir.

Captain J.J.: They're everywhere.

Dulcinea: I think this is it.

Amadeus: We don't have a chance at all.

Simbra was crying.

Dulcinea: (alien language) It's okay, it's okay.

They were in a circle again, but they were surrounded by the Water Knights. By that time only a miracle could save our heroes. So it was one of too many scary moments. Our heroes were in deep trouble and nothing seemed possible to rescue them. They didn't have backup.

Dulcinea: They're getting closer.

Amadeus: I know.

Captain J.J.: Come on, I'm waiting, come on.

A Water Knight was going to attack.

Dulcinea (holding Simbra in her arms): Here they come.

A ray of various colors showed up from nowhere, targeting every Water Knight. Instead of exploding into water, the Water Knights were frozen; then they melted and then vaporized. It was the only way to get rid of those warriors made of water, but who came to save them?

Dulcinea: What's that?

Amadeus: We're safe.

Dulcinea: Yeah, but who saved us?

Amadeus: I don't care, we're just safe.

Dulcinea: I don't see anybody.

Captain J.J.: There, somebody is coming.

A group of people arrived.

Captain J.J.: There, it can't be.

They were the Chameleon's crew, the other ship that was thrown miles away when they were trying to get onto Europa's surface.

Jenss (in a German accent): Hey, are you okay?

Dulcinea: Definitely, you guys were just in time.

Amadeus: But where have you been?

Jenss: We got lost. Basically we were lucky, we survived the crash; some of us didn't make it.

Dulcinea: So how many survived?

Jenss: Just a few, fifteen in all.

Amadeus: From how many?

Jenss: Fourteen hundred and fifty people, not including the officers.

Captain J.J.: So who's in charge?

Jenss: Nobody, we make decisions in voting, sir.

Captain J.J.: So how about that weapon? Where did you get it?

Jenss: We got it from here, sir.

Captain J.J.: Be more specific.

Jenss: We got it from him.

He pointed at a guy around 40 years of age. On the other side, Lt. Dago and around ten soldiers were coming too.

Captain J.J.: A civilian, you were bringing civilians to this mission?

Jenss: He's not a civilian, well, he's a civilian from this place.

Dulcinea: Yeah, he's from here.

Jenss: How did you know?

Dulcinea: Sir, let me speak to him.

Captain J.J.: Go ahead.

Jenss: It's pointless, he doesn't speak English.

Amadeus: She's not going to speak English.

Captain J.J.: Ask him about the weapon.

The weapon was like a Roman shield but smaller; it was equipped with some kind of gun in the middle of it, from which the rays of color came out.

Dulcinea: (alien language) Hi, my name is Dulcinea; we're from planet Earth and we're here to help you.

Tamuz: (alien language) Greetings, my name is Tamuz, and I'm the leader of the Bacan tribe.

Dulcinea: (alien language) We're here to defeat the Zurdos, so you must be the rebels trying to break free from Europa.

Tamuz: (alien language) Yes, we've been fighting for such a long time that it seems this war is useless. We've lost a lot of our kind.

Dulcinea: (alien language) Don't worry, we're here to join forces. We have something in common, to destroy the Zurdos.

Tamuz: (alien language) It's really hard to defeat the Zurdos; we've been running away for so long that there's nothing that can save us.

Dulcinea: Don't lose your faith, we're in this together. Show me where the rest of your people are.

Tamuz: I'm the only one. I was lucky to escape. They destroyed my village and the survivors were taken prisoner for experiments; they do horrible things to our people. I'm glad that you are going to help us.

Dulcinea: How about that weapon?

Tamuz: Our scientists made this unique weapon from Super Eurotanius to defend us from the Zurdos, especially from the Water Knights.

Amadeus was translating from the alien language to English.

Dulcinea: I wouldn't worry about these Water Knights anymore, they're gone.

Tamuz: There are more powerful and destructive commanders from the Zurdos' empire. They're settled in the Death Valley.

Death Valley was where Alexandrus was headed.

Dulcinea: There are more challenges, right?

Lt. Dago: Are there more powerful warriors than these Water Knights?

Amadeus: Don't forget destructive also.

Lt. Dago: Oh my God.

Captain J.J.: Don't interrupt, Lt. Dago, we need to know more.

Tamuz: Even with your people, it's not enough to destroy Atonal. He's so powerful, and his army is too much for us to handle.

Dulcinea: Well, we'll find a way to destroy him.

Captain J.J.: Ask him about the big city.

Dulcinea: Where's Atonal, is he in the big city?

Tamuz: You mean the City of Fury?

Dulcinea: Yeah, the City of Fury.

Tamuz: I'll take you there. We'll be there in two or three days.

Dulcinea: Do you have more of those powerful shields?

Tamuz: Well, our tribe used to have some, but the Zurdos knew it. They thought we were a threat to them. This one I saved from them; I buried it. It can be very useful, but believe me, there's more danger out there that this is not going to work against.

Dulcinea: So they decided to wipe you out.

Captain J.J.: Hey, Dulcinea, ask him if there's any way to get there faster.

Dulcinea: Is there any way to get there faster?

Tamuz: Before the "Big Hunt" our place was so advanced, we had everything. We used to tame the air, water, and the underground, but now it's all gone. We live in fear. There's no future for us—please help us.

Dulcinea: How about food, weapons? We need more supplies.

Tamuz: Well, not far from here, there is another village, Tazumal. We have to get there first. It's the only place where we might find food and something to fight with, but there's a little problem.

Dulcinea: What is it?

Tamuz: It's under the Zurdos' control.

Dulcinea: Well, let's go to Tazumal.

Captain J.J.: Okay, listen up, people, this is the plan. We are going to go to Tazumal; probably we'll find food, weapons, or nothing, but we have to get there.

Lt. Dago (softly): Sir, do you think we might find the Super Eurotanius in Tazumal?

Captain J.J.: We'll see, Lieutenant, we'll see. We need to regroup—we're 50 in all.

Right away, they started to move. Every time they were battling for survival, they were getting smaller. Dulcinea was with Simbra, Amadeus was with this new ally, Tamuz, Captain J.J. was with Lt. Dago, and Jenss was with his partners. So they were walking with high expectations with this new challenge, Tazumal.

Alexandrus, Colonel Demello, and the rest of the crew were still on their own journey. They'd crossed beautiful mountains, rivers, and after a long walk they stopped.

Alexandrus: Wow, this place is beautiful.

Colonel Demello: It's a shame those Zurdos are trying to destroy it.

After a couple of minutes admiring the place, Alexandrus noted something.

Alexandrus: Sir, did you see over there?

He pointed at the top of the hill.

Colonel Demello (grabbing his binoculars): Let me see. Yeah, you're right, there's something over there. Wow, you have good eyes.

Alexandrus: I will ask Miche about that spot.

Colonel Demello: Okay, go ahead.

Alexandrus went over to Miche.

Alexandrus: (alien language) Hey, Miche, do you know if there's something around there?

He pointed.

Miche: (alien language) That's Izalco, a small city with brilliant people.

Alexandrus: Any Zurdos at all?

Miche: I don't know, you might find Zurdos anywhere.

Alexandrus: Well, we'll find out soon.

And our heroes went there, like always, ready for anything. They knew that they might be facing a confrontation. On the frontlines were Alexandrus, Colonel Demello, Atlacatl, and Miche, and they were getting closer and closer.

Alexandrus: Sir, we should be careful, it might be a trap.

Colonel Demello: It might be.

When they got there, the scenery left them speechless. What our heroes were witnessing was scary, terrible. The fog and the smog were the only ones who witnessed what happened to Izalco; their village was completely wiped out, and there were many dead people lying on the ground. There were no survivors. Something merciless had destroyed everything. Those people didn't have a chance at all to defend their lives. It was like a cemetery on the road. Our people were searching for survivors but without any luck.

Alexandrus: What the hell happened here? It's horrible.

Atlacatl: There are no survivors, everybody is dead.

Alexandrus: These people didn't have a chance to defend themselves, look at them.

Colonel Demello (running up): Hey, Alexandrus, there are no survivors. Something horrible happened here.

Alexandrus: I know, sir; but there's something that I don't like.

Colonel Demello: What?

Alexandrus: It doesn't look like they were attacked by foot troops.

Colonel Demello: What do you mean?

Alexandrus: They would have had a chance to defend this place.

Colonel Demello: So what attacked these people?

Alexandrus: I don't know, sir, I don't know, but if you see the way that they were killed, it's strange. It seemed that something big struck them.

Colonel Demello: A kind of giant warrior.

Atlacatl: Or something from the sky.

Colonel Demello: It could be that they were poisoned by an airborne virus.

Alexandrus: I'm not sure, because there's blood on the ground.

The troops were still looking for survivors and the shacks were still burning.

Alexandrus: But this attack just happened.

Atlacatl: Hey, what's this?

He had found something like a small computer on the ground.

Alexandrus: Probably it's like a kind of message or like a black box.

They looked around it, trying to turn it on, and then they found a way to play it.

Woman's voice in alien language: Whoever finds this message (blurry image), we were supposed to get much better protection against the Zurdos. We knew the risks. (blurry) I have to be very quick, they're coming.

She turned around to look over her shoulder. There was something coming.

Woman's voice: They have killed most of us; the alliance has to know this—they took the antidote. Please forgive us, we failed. If someone (blurry) somebody has to speak out for these people.

It was gone.

Colonel Demello: What happened here, Alexandrus?

Alexandrus: The extermination of a tribe; they were trying to make their own race. Something went wrong. If this attack just happened, look around. Why do they still have their artillery in position? They didn't run and they weren't scared.

Atlacatl: It could be the Topos.

Alexandrus: No, they're something much stronger. I think someone should know something. Hey, Miche, any idea what happened here?

Miche: I'm not sure, but it could be anybody from the Zurdos.

Alexandrus: Well, that I'm sure of, but who? Any commander in particular? How about Atonal?

Miche: Well, for sure he didn't do it—he has his evil servants doing his dirty work, and before you ask me, yes, he's the strongest one from all the Zurdos Empire.

Alexandrus: Ah, and is there any weakness about Atonal?

Miche: Well, he seems invincible in this place; the few who have dared to go against him didn't make it.

Alexandrus: Have you seen him fighting? What makes him so great?

Miche: Listen, we're not allowed to go to his Chamber, just the commanders, but not so long ago he was challenged by one of his commander's troops, Cuzcatlan. He and his troops were exterminated in minutes just by Atonal—his speed, his strength, his weapons made him a very deadly force. After he won the battle, he displayed Cuzcatlan's body and the rest of his battalion in front of his palace. He also has a lot of supporters, and they were cheering him.

Alexandrus: Why did he challenge him? That means that there could be others who might do the same but they are scared.

Miche: Because not all of us agree with his thoughts. He's a killer; he's so bad that he doesn't care about this place. He's a conqueror. He destroyed Atlantis in the red planet, he's crazy in destroying your place, the Blue Land, and you're right, there might be others like me out there waiting.

Alexandrus: Here, my friend, we're the ones. We're not going to allow them to destroy our beautiful Earth, the "Blue Land" as you called it. We are going to help you and the rest of your people in Europa.

Miche: Thanks a lot, but it's going to be hard. I still remember the speech that he gave us before you guys landed. He was expecting you for so long, he called the 13 tribes just to take care of all you. The priest from the 13 tribes was calling his guards from the ocean's darkness. They are bloodthirsty killers, and he was calling all the tribes to get united to destroy the Blue Land. Atonal said, "I'm imploring all of you to unite forces to defeat this common enemy; they have come to enslave us. We were born free and our ancestors had kept the invaders from this land. We will not tolerate these invaders taking our blessed soil. This is ours, yeah, this is ours… I'm not asking citizens from Europa for help, I'm not begging you citizens from our sacred land… I'm commanding all of you to join this fight against the Blue Land. Death to the Blue Land, death to the Blue Land!" All his people were chanting his call.

Alexandrus: I know, we don't take anything for granted. We have to fight for it. What about the Super Eurotanius? Is it here?

Miche: It's in Atonal's city.

Alexandrus: The City of Fury?

Miche: With the Super Eurotanius, Atonal is planning to conquer your planet and the rest of this galaxy.

Alexandrus: Not if we act first. I'm going to go find Colonel Demello. Hey, thanks for all that information.

Miche: You're welcome.

Not far away, Colonel Demello was alone, smoking a Cuban cigarette and admiring the scenery—the landscape of Europa. He also was nervous. At that moment the sergeant came up to give a full report.

Sergeant: Sorry, sir, for interrupting you, but we've searched the whole area, and there are no survivors. We're also looking for the Super Eurotanius, but there's nothing here. Professor Kishimoto and his colleagues are still on the track, sir.

Colonel Demello: Okay, well done, Sergeant, you're doing a good job.

Sergeant: Thank you, sir. We also found food and some weapons like swords.

Colonel Demello: Take whatever you think we might need, and remember that we're almost out of food and bullets.

Sergeant: These kinds of swords might help us, sir; we've seen that they are more useful in combat, especially when there's fog.

Colonel Demello: And you're right, Sergeant, they were more effective in the last combat. Take them all, go.

Sergeant: Yes sir.

Colonel Demello bent over to tie his boots and noticed something weird. The sergeant was still there, right behind him.

Colonel Demello: (still tying his boots) Sergeant, what happened? Why are you still here?

Sergeant: Oh my God, what's that coming?

Colonel Demello: Where, Sergeant?

Sergeant: (pointing at the sky) Over there, sir.

Colonel Demello: (looking up) Jesus, what's that?

Colonel Demello's Cuban cigarette fell from his mouth. They were freaked out by what they were seeing in the sky. There were many black points on the horizon, a big dark cloud of them, and they were approaching. As they got closer, their shapes were changing gradually, and they weren't black points anymore. They were transforming into humans, yes, humans with wings—the enigmatic Aguilus, a kind of invisible eagle men.

Colonel Demello: What the hell is that?

Sergeant: Let's get out of here, sir.

Colonel Demello: You betcha, Sergeant.

They stared to run…and very fast. The Aguilus were the result of an experiment by the Zurdos between birds and civilians from Europa.

They were one of the strongest guards from Atonal's Empire. One of their unique skills was invisibility, and they could fly, walk, run, and carry swords. They were strong and merciless too. When these eagle men were attacking, they didn't leave anybody alive. They would be a difficult challenge for our heroes.

Alexandrus: What's that?

Atlacatl: Angels? They're coming.

Alexandrus: Listen, everybody, we're under attack! Run fast!

Colonel Demello: Any idea what we're dealing with?

Alexandrus: No sir, let's hide quickly.

Miche: Oh no, they've sent their best killer warriors.

Alexandrus (running alongside Miche): Do you know them? What are they?

Miche: We're dead now.

They were running; everybody was trying to get away from this new enemy, but there were some soldiers standing, firing their guns, and missing their target because the Aguilus were disappearing in seconds. Then they were killing the soldiers. That attack was a big surprise to our heroes; they didn't have the time to counterattack. In minutes, those Zurdos had taken a lot of soldiers down. Their swords were too much against our heroes. There was chaos, there was no escape—just waiting for a quick death. Alexandrus was running away and he found a kind of pile filled with the swords that they found earlier.

Alexandrus: I'm so lucky. I can't believe it.

He didn't have much time to run, so he stood up, grabbed a sword, and just waited for his target. The eagle men were invisible except when they became visible to destroy their enemies. Alexandrus knew right away that he might have a chance to kill those flying monsters.

Alexandrus: Aha, I gotcha.

He was steady, looking around. He knew that sooner or later his turn would come to see his enemy.

Alexandrus (to himself): Come on, where are you?

In a blink of an eye an invisible eagle man appeared behind him, ready to kill him. Alexandrus turned around; he knew that his rival was behind him—he just needed that short time to focus on him.

Alexandrus: I got you, yes.

He pierced his adversary's chest with his sword, and it resulted in a big victory for Alexandrus. He needed to be ready for that short period of time again when the Aguilus were vulnerable. One down, twenty-five to go.

Alexandrus: So, you're not invisible after all. Take this from planet Earth.

He went to look for another one, and again, he succeeded.

But Alexandrus was not the only one doing his part. Atlacatl was using the same technique.

Alexandrus: Well done, Atlacatl, like the training we had before.

Atlacatl: Yes, be careful, there are more coming.

The Aguilus were killing on their side too, but they realized that two great warriors were giving them a hard time; those two warriors were also killing them. Colonel Demello was astonished; he couldn't believe what his eyes were seeing—those two young warriors against one of Atonal's fiercest troopers. He and Professor Kishimoto were hiding but witnessing the fight.

Colonel Demello: I'm glad that we're on their side.

Professor Kishimoto: They're extraordinary. Without them we would be history by now.

Colonel Demello: Look, Atlacatl's platoon is joining too. We should help them.

Professor Kishimoto: But how? They're really fast and they're flying.

Colonel Demello: You're right but I have to help them.

Alexandrus and Atlacatl were really great, but they were humans too; they were getting tired, and all the Aguilus were focusing on them.

Alexandrus: There are more coming.

Atlacatl: I know, we need backup. My soldiers are coming.

At that moment Atlacatl was knocked out by a hard punch.

Alexandrus: Are you all right, Atlacatl?

There was no reply. Atlacatl's platoon was fighting with their swords too.

Alexandrus: Get up, Atlacatl, get up!

Colonel Demello acted right away; he was firing his gun, of course, at anything in the air, close to our heroes but without any luck. He was trying to get their attention. The Aguilus were trying everything against our heroes; some of Atlacatl's platoon were killing them, but they were also getting killed too. They were wounded and tired. There were many dead soldiers on the ground, and the few standing were just waiting for the final moment of their lives. Then there was a kind of blast or explosion in the sky that turned part of the sky a reddish color. It was spreading some kind of substance around the air in the battle, and that substance was reaching everything in its way. It turned into a "red spot," which meant that anytime the substance touched the Aguilus, they became visible. That changed everything. Now they could be spotted. The battle began all over again, but this time it was even.

Alexandrus: What?

He was barely standing, bleeding from one of his arms.

Colonel Demello: What's going on?

Surprisingly there was a rain of arrows flying towards the Aguilus. The enemies were falling down one by one; someone was saving our heroes, who were dominated by fatigue. The few Aguilus that were left were getting away.

Alexandrus: They're running but from who?

Colonel Demello and the survivors came running toward Alexandrus.

Colonel Demello: Did you see that, son?

Alexandrus: Yes, I saw that, but who is helping us?

Atlacatl regained consciousness.

Atlacatl: What happened? Did we win or what?

Alexandrus: Yes, but we'll find out soon who's helping us.

Colonel Demello: Over there.

Everybody looked where the colonel was pointing. There was a group of about fifteen people coming at them, carrying swords and arrows, and some of them were carrying some kind of guns. They were the rebels or Alliance from Europa.

Atlacatl: Are they friendly?

Alexandrus: I think so or else they wouldn't have helped us.

Miche: They're the rebels.

Alexandrus: (alien language) Greetings from planet Earth, I mean the Blue Land. My name is Alexandrus and we're here to help you.

Shan: (alien language) Greetings from my people. So you must be the ones that our people are talking about. We appreciate your help. Now I know that our messengers got to their destination.

Alexandrus: Yeah, and one of them was my father.

Shan: My name is Shan, and I'm the leader of the 13th tribe; we have been fighting the Zurdos since the big attack on the Red Land.

Alexandrus: Atlantis.

Shan: Aha, well, I guess you're ahead in this.

Alexandrus: Not really, we have one of your guys, Miche. He's been telling us about the whole thing.

Miche (arriving): So you're Shan, leader of the 13th tribe. Atonal is so obsessed with your head; he has sent his troops after you and wants you dead or alive. Well, I'm honored to be on your side; together we might have a chance to destroy Atonal.

Shan: The honor is mine, to be in this together, and you're right, if we unite our powers, we can defeat Atonal.

Alexandrus: Yes, this is why we're here.

Shan: Let's go. We'll take you to the camp. The Zurdos know that we're here, so we should leave immediately. Follow us.

Alexandrus: Okay, I'll tell my superior.

He turned to Colonel Demello.

Alexandrus: Colonel, we've met the rebels, and these people need our help. We're going to their base, and we should leave right away— the Zurdos might come. I'll tell you the rest on our way to their base.

Colonel Demello: Okay, Alexandrus, but before we leave, let me have a moment of silence for our departed soldiers.

Nearly 70 percent of the soldiers were dead.

Alexandrus: Colonel Demello, you're right. They made a big sacrifice for us.

Professor Kishimoto (arriving): Sir, Sergeant Smith is dead, we have about 15 on their feet.

Colonel Demello: Oh my God, we're so few; we're never going to defeat the Zurdos.

Alexandrus: But we have to finish this mission, people. Earth is praying that we destroy the Zurdos, and thank God, we have new allies.

Colonel Demello: Okay, let's pray for them, and let's get out of here.

Alexandrus: Yes, Colonel.

They had a minute of silence for the dead soldiers killed in the battle. They were very devastated by this war, and they were without any hope, but they had to finish it. They were following Miche, Shan, and his people. Alexandrus was walking with Colonel Demello.

Colonel Demello: Son, it's been a long time since "the Landing" and the things we went through. We have lost a lot of people in this mission, and I want you to be honest, what do you think?

Alexandrus: To be honest, since I left planet Earth, I knew that this mission might be a one-way trip, but seeing all the faces of our people, my family and friends, I will never let those people down, even if it's the last thing I do! Today we had a small victory, thank God. We were lucky this time, but we're still far from the finish line, and we will not accept defeat, especially after all the sacrifices that our people have already made. We're going to finish no matter what; we'll do it for planet Earth, for our people, our families, and most for all, our future generations, our children.

Dulcinea, Amadeus, and their group had found a river, and were trying to decide whether to cross it by boat or walk around it—of course the last choice would be longer.

Lt. Dago: Sir, there's a river, what should we do?

Captain J.J.: We'll cross it.

Lt. Dago: But we could be an easy target.

Tamuz was searching some bushes in which there were some boats hidden.

Captain J.J. (pointing at Tamuz): Over there. That's your answer, Lt. Dago.

Lt. Dago: With all due respect, I think we shouldn't...

Captain J.J.: Do as I said, Lieutenant. I don't want to repeat it again, understood?

Lt. Dago (angrily): Understood, sir.

He left. Dulcinea came up with Amadeus.

Dulcinea: Sir, you're making a big mistake. This river might be dangerous.

Captain J.J.: I don't make mistakes, woman. Are you telling me what to do?

Dulcinea: Captain, we don't know the river. Lt. Dago is right, we might be an easy target. Look around, it's very quiet.

Captain J.J.: I'll have you court-martialed when we get back to Earth.

Dulcinea: You don't see, everybody is scared. Your decision might cost us the whole mission. I'm sensing that there's something out there, and we're not going to make it. We're not ready to cross it. We should walk around it. It will take more time, but it would be much safer and maybe we could find Colonel Demello and the others.

Captain J.J.: I'll give you three seconds, or I'll take you personally to prison.

Amadeus: No, sir. it's okay, we're leaving.

He grabbed Dulcinea and they left.

Dulcinea: What are you saying?

Amadeus: It's okay, he's right, we should accept his orders.

Dulcinea: But you know I'm right.

Amadeus: I know. I'm sensing danger in that river too.

Dulcinea: And are we going to let that happen?

Amadeus: We don't have any choice, he's in command.

Dulcinea: We should do something.

Amadeus: There's nothing that we can do.

Dulcinea: You're right, but Colonel Demello has to know this.

Amadeus: Yeah, he will know, believe me.

Angrily, they went to the river's shore; everybody was there, preparing the boats. They were ready to cross it, but...

Dulcinea: It's okay, Simbra, we're going to the other side.

Simbra: I don't want to get on those boats.

Dulcinea: Nothing is going to happen, come on.

Simbra: You promise me that you're going to stay with me?

Dulcinea: I promise. Come on, I'll protect you.

Simbra and Dulcinea got in one of the boats; Amadeus was in

another boat, and Captain J.J. was escorted by more soldiers than the rest. There were around fifty in all, and they had to split the groups. There were five boats, all of them trying to cross the river. The first boat was full of soldiers, the second carried Captain J.J. and Lt. Dago, in the third was Dulcinea, in the fourth was Amadeus, and the last one was carrying the rest of the soldiers. They're in the middle of the river.

Lt. Dago: Sir, we're almost there.

Captain J.J. didn't answer. He was serious the whole way, just looking around. Meanwhile, in the third boat.

Dulcinea (pointing at her side): Hey Tamuz, what's over there?

Tamuz: It's the only road to get us to Atonal's city.

Dulcinea: Any chance to beat him?

Tamuz: Honestly?

Dulcinea: Yeah.

Tamuz: I doubt it, but…

Dulcinea: What?

Tamuz: When there's hope, there's a chance.

There was a complete silence in the water, and our heroes were close to their destination. Everybody was quiet and ready for any action. All of sudden there was a snorkeling sound and air bubbles in the water, and everybody was confused. The soldiers were pointing their guns into the water, and what happened next was the captain's biggest mistake. The first boat was assaulted. The Zurdos, in an unusual attack, destroyed the first boat completely; they were well hidden under water. And then the rest of the boats were under attack.

Lt. Dago: Sir, they caught us, we're under attack.

The soldiers were firing their guns, shooting at the river, but they couldn't see the target at all.

Dulcinea: It's okay, Simbra, just stay close to me.

It was a big battle, and somehow, the soldiers were hitting their targets; there were a few Zurdos floating in the water. When they

thought that they were in control of the situation, the real threat showed up. The fog was closing on them, and there was something strange swimming around the captain's boat. It was fast, and it didn't look like another of Zurdo's soldiers.

Lt. Dago: Sir, did you see that?

Captain J.J.: Yes, what was that?

Lt. Dago: I don't know, sir, but it looked like a giant crab.

Captain J.J.: Don't say stupid things.

Lt. Dago: Sir, here it comes again.

Captain J.J.: Open fire.

They opened fire, but that thing was really fast. Sometimes they could see it and sometimes they couldn't, but this time that "giant crab" took two sharp swords. It jumped into the air and killed two soldiers right away, then disappeared.

Lt. Dago: What was that thing?

They were attacked again by this really fast creature.

Captain J.J.: Don't waste the bullets, just be in position.

And once again that thing was coming right at them. It jumped into the boat, killed some soldiers, but this time it didn't get away. It stood up in one of the corners of the boat, holding one soldier's head with its arm, and Captain J.J., the lieutenant, and five more soldiers were witnessing this monster—"Cangre," another of Atonal's killers. It was in command in the water, master of Europa's ocean. Its skin was very unusual—a combination of a crab and a human—and it could breathe in the water.

Lt. Dago: What in God's name is that thing?

Captain J.J.: Lieutenant, we better get out of here.

Cangre took two soldiers with him into the river and then they got away. There was a lot of blood in the water.

Lt. Dago: He vanished.

Their boat was flipped over, and everybody jumped out, swimming and trying to reach the shore.

Lt. Dago: Sir, over there.

He pointed at the other side of the river. Then Dulcinea's boat flipped and everybody fell into the water. Amadeus was in the boat behind Dulcinea's.

Amadeus: Hey, Dulcinea, hold on!

He jumped into the river, carrying his sword and two knives. The soldiers from Amadeus's boat were trying to help the others and were also fighting the other Zurdos soldiers.

Amadeus (telepathically): Where are you, Dulcinea?

Dulcinea (floating in the water): I'm here, get the others.

Dulcinea was trying to reach Simbra when she saw Cangre coming.

Dulcinea: Oh oh. What's that?

Cangre attacked Dulcinea with his sword, but Dulcinea was really fast and got away. She was lucky this time, and she put her sword right in front of Cangre's two swords, but she lost it—her sword was broken in two pieces. Then again Cangre was coming at her, and this time she was helpless, swimming for her life. She was trying to reach the shore when she saw him coming.

Dulcinea (to herself): I've got to hurry up.

Cangre was getting closer and closer and then he reached her. He lifted one of his swords and was going to kill Dulcinea when another miracle happened. From behind her something stopped Cangre's sword from reaching its target; it was Amadeus's sword holding back a ferocious blow from the enemy. They saw each other, knowing that it was going to start another epic battle. Dulcinea went right away to look for Simbra. The only boat left was picking up the survivors. Amadeus and Cangre began to fight.

Amadeus: Whatever you are, you're going to pay for this.

With his hand holding his sword, Amadeus made the first move, but Cangre didn't wait either; he was using both swords. They were focusing on each other with good reflexes, fighting sword by sword. Amadeus thought that he had the victory, but when he was going to finish him off, the unexpected happened again.

Amadeus: All right, it's time that you pay for all these unforgivable attacks…yahhh!

Suddenly Cangre was starting to change or mutate; he was turning into a ball like a big water mine, ready to make contact and explode.

Amadeus: What?

Cangre was coming at him with an incredible velocity—no chance to get away.

Amadeus: Oh my God, it's coming right at me.

Amadeus's only chance to escape was to try not to be in his way. When Cangre was very close, Amadeus took a deep breath and submerged, so this time "Mine-Man" missed his attack but not before destroying a flipped boat. Mine-Man was looking around—no sign of Amadeus; the rest of the soldiers and the boat were getting close to the shore, and it was impossible that Amadeus was over there, so Mine-Man was still looking for Amadeus when he saw something swimming to the shore. He turned back into a ball and went after that spot. He was really scary on the water. Amadeus was trying to escape. He struck it with fury—no chance of survival after that attack. Mine-Man went to check it out right away; he thought that he had killed Amadeus. The body was floating and when he grabbed the body to see his face, he became very angry.

Mine-Man: ARRGHH.

It wasn't Amadeus's body, it was a soldier trying to escape. Amadeus counterattacked and cut part of Mine-Man's shield. The attack was in vain.

Amadeus: What? What are you made of?

The shield had protected Mine-Man's body; it was almost the same as a crab's shell. Cangre was angry. He grabbed Amadeus and they submerged. Cangre's tactic was to drown Amadeus, but he didn't know that Amadeus was a good swimmer back on Earth, so they were battling again, but this time under the water. Mine-Man was holding Amadeus, who was trying to break free from that "mortal grip." It seemed that Mine-Man's strategy was not to give him a break in the water. Amadeus was doing his best but without any luck; Mine-Man was drowning him. Amadeus needed oxygen, and he needed to do something and quick. In his last effort, Amadeus grabbed his knife (which was under his uniform's belt) and knocked his head against Mine-Man's face; then he put his knife in Mine-Man's chest. It gave him enough time to get away; he couldn't hold his breath anymore. He had to go to the surface. With a lot of trouble he almost drowned under the water, but he reached the surface and he was swimming to the shore. He didn't care to know if he had successfully killed the Mine-Man under the water; he just wanted fresh air to restore his energy. He was alone, the last one to reach the shore. He was so tired when he touched the ground he fell down, almost unconscious.

Amadeus: Thank you, God, you saved me again.

Everything seemed okay. He couldn't see the rest of his people so he just wanted to take a break and then he would look for the rest of his friends.

Amadeus (exhausted): I'm tired.

Then he heard some footsteps getting close to him and decided to turn his head. All of sudden he heard something breaking the silence of the air, and he moved quickly. Mine-Man's sword crashed in the place where Amadeus was. It was Cangre. He was bleeding and he started to attack with his sword. Amadeus had lost his sword in the water, and he was defenseless. He was just barely avoiding Mine-Man's sword. He was running away from his adversary, but luck was on his side. When he had no chance to get away he found a sharpened piece of wood from a wrecked boat and that was what he needed. Mine-Man was very angry; he launched the first move and with extreme caution, Amadeus held the Mine-Man's fist, which held the sword, and pushed with all his strength the sharpened piece of wood inside his enemy's chest. This time he was killed for sure; there was no hope that he could survive such an attack.

Amadeus: Take this for all you have done.

Cangre fell slowly, dead by the time he hit the ground. Amadeus had won. Quickly he decided to leave; he needed to find his friends, but not before taking his enemy's weapon.

Amadeus: I'll take this. I think I'll need it more. Thank you.

He left but he didn't get far because Dulcinea, Simbra, Lt. Dago, and Captain J.J. were close by.

Dulcinea: Hey, Amadeus, over here.

Amadeus (waving): I'm here.

Lt. Dago: What happened?

Amadeus: It's over, we have won this battle.

Dulcinea: We have lost almost everybody.

Amadeus: How many?

Lt. Dago: There are eleven left, including you.

Amadeus: Oh my God, we're just a few, we're never going to make it.

Dulcinea: None of this would have happened, right, sir?

Captain J.J. didn't answer; he just ignored them and left.

Amadeus: What's wrong with him?

Lt. Dago: Dulcinea was right, he feels guilty about this.

Dulcinea: They almost killed us. We don't have weapons, food, medicine. We're just eleven and we're lost.

Amadeus: Calm down, Dulcinea, just relax, we need to regroup and set some strategies.

Dulcinea: You're right.

Amadeus: Hey, what about Tamuz? Is he alive?

Dulcinea (sobbing): No, he didn't make it.

Lt. Dago: Hey, I want to tell you one thing—the others don't want to follow the captain's orders; they want to follow you guys. They think Captain J.J. is out of his mind.

Dulcinea: Well, it's a privilege to hear that, but we can't do it.

Lt. Dago: What do you mean?

Amadeus: Well, in that case, it's you, Lt. Dago; if something happens to Captain J.J. you're second in command, so you're in command.

Lt. Dago: Yeah, you're right.

Dulcinea: And remember, you have to win the respect of your soldiers.

Lt. Dago: You're right.

Dulcinea: Okay, people, we should go right now.

Amadeus: I'm hungry and tired; we should look for some food and shelter.

Dulcinea: Me too, let's find a place to rest.

Lt. Dago: I'll check with Captain J.J.

Dulcinea: All right, let's get out of here before it gets dark.

The shore was tainted with blood. It also started to rain very hard, with strong winds. They were survivors and they were lost. Finally they found a good spot for resting, and they were proposing strategies to follow and eating the last cans of food that they had. They weren't ready for another attack; they were vulnerable. Everybody formed in a small circle, giving details about their lives and sharing (probably) the last minutes of their lives. It had been a couple hours since the attack, but it seemed that daylight wasn't coming soon. Someone was missing in that group, someone who should have been there, someone to take the blame and responsibilities. It was Captain J.J. missing the meeting; he was outside of the group. He was so ashamed of what had happened earlier in the river. Then one by one the others went to sleep, and he was the only one awake; it was his turn to watch the others, but he didn't say a word.

Meanwhile not far from there, Alexandrus, Colonel Demello, Atlacatl, Miche, and the others were still fascinated by those rebels. They were taken to their base, and by the time they reached their destination it was the middle of the night. They were also very exhausted.

Colonel Demello: This must be their place.

Alexandrus: Yeah.

Atlacatl: And they're having a party?

Alexandrus: It looks like it.

Alexandrus: What's going on, Shan? Are you having a party?

Shan: Yes we're making a sacrifice to our god, Chimbe.

Alexandrus: What?

Atlacatl: What happened?

Colonel Demello: Yes, tell us what's going on?

Shan: It's our ritual; we venerate Chimbe for all the harvest in the season.

Alexandrus: Sir, I don't know what to say; I'm not even sure if we can trust these people.

Colonel Demello: Why? Tell us.

Alexandrus: They're making a sacrifice for some god called "Chimbe." They think he's responsible for this harvest each season.

Professor Kishimoto: Interesting, the lifestyle of these people; they still believe in different gods.

Shan: Would you do us the honor of being present in our ceremony?

Alexandrus: Sir, I think you're not going to like what I'm going to say.

Colonel Demello: What? Tell us.

Alexandrus: They want us to be present at a sacrifice that they're going to do.

Colonel Demello: Are they insane?

Alexandrus: I don't know, sir.

Atlacatl: We probably have it all wrong; maybe they're having a different sacrifice.

Alexandrus: I don't know, sir, but I'm assuming that wagon is not empty.

Atlacatl: You're right, there's someone in there.

Alexandrus: Just wait. Hey, Shan, who's inside of that wagon?

Shan: She's the chosen one. Come on, I'll introduce you to our Cacique.

Alexandrus: Cacique? Wait, I've got to tell this to my superior. (To Colonel Demello) Sir, it's their leader, or whoever, they call "Cacique" and they are going to take us with him.

Colonel Demello: Okay, maybe we can change his mind.

Alexandrus: I don't think so, sir; they seem to have their customs, and we can't just change them overnight!

They were taken to Cacique and were confused by what they were seeing. They were watching some kind of dance around the fire, and the music was coming from some rustic instruments like flutes, drums, and other rustic objects. Their "camp" didn't look settled because they had to move constantly from the Zurdos. It was like a temporary place for them.

Shan: Wait here; I'll go see him first.

Shan went to Cacique and our heroes were waiting.

Alexandrus: I hope this doesn't change everything.

Colonel Demello: I think it will.

Professor Kishimoto: Hey, guys, this is amazing.

Alexandrus: What are you saying?

Professor Kishimoto: You don't have any idea? These people could be like our ancestors; of course they have evolved their weapons, technology, medicine, clothes, and maybe they were ahead of us about thirty years, but their culture, their customs are the same.

Colonel Demello: Professor, speak English, be more specific.

Alexandrus: I got ya, Professor. Sir, what Professor Kishimoto is trying to say is that these people have the same customs, culture, and religion as the Aztecs, Mayans, and Incas before Christopher Columbus discovered America.

Colonel Demello: I got ya.

Professor Kishimoto: Yeah, that means that if we weren't discovered by Columbus, we would've developed such technology on our planet.

Alexandrus: But remember, we would've had humans sacrifices too.

Professor Kishimoto: Yeah, but we would've lived in a perfect society. I mean, these people were ahead of us by many years before the war of course, but still we can learn a lot from these people.

Alexandrus: Well, we can learn from them, I agree, but no sacrifices.

Shan was coming with Cacique.

Shan: This is our Cacique; he's the leader of all the tribes left in this place.

Chafa (alien language): Greetings, strangers, my name is Chafa. I'm the leader of this tribe. I've been told that you came from the Blue Land and you're going to help us destroy Atonal.

Alexandrus (alien language): Mmm, my name is Alexandrus and we came from planet Earth—I guess that's the "Blue Land" for you guys. We're here to destroy Atonal and all his regiment.

Chafa: Where's the rest of your army?

Alexandrus: Well, the rest of our army is…hold on. (to Colonel Demello) Sir, they're asking where our army is.

Colonel Demello: Tell him that our backup is coming with powerful weapons; we just want to be sure that Atonal is gone.

Alexandrus: Sir, I don't think that they're going to buy it.

Colonel Demello: Any better ideas?

Alexandrus: Not now, but we should tell them the truth.

Colonel Demello: And what if these people are not telling the truth? We don't know them.

Alexandrus: Sir, they saved our butts out there.

Colonel Demello: Okay, just tell them that our backup is coming.

Alexandrus (to Chafa): Sorry for this but I had to speak with my boss. Well, the thing is that the rest of our army is on their way to Atonal's city. We've lost a lot of soldiers in our mission and we had to split in two groups; we don't know anything about the other ones, but I'm sure they will be fine. And there's more. We're out of food, ammo, and we need supplies. We need to get Atonal's city no matter what—we're not going to let those bastards destroy our planet if it's the last thing we do.

Chafa: I'm grateful. What you're saying is the truth. Before the war we used to be friends with the Zurdos. As you know we used to live on Jupite, but destiny sent us a big shower of asteroids; we asked Atlantis for help but they refused, and we were almost wiped out from the asteroid showers. We came here, we started all over again, and we discovered this super mineral. It was like a gift, and at the beginning it was great. We built great cities, great weapons, great industries, and our new home, but we lost our way. We started to do experiments on ourselves. The Zurdos were always trying to take advantage, and they built a great weapon that destroyed the Red Land. It was like payback to the Atlantians for not helping our ancestors.

Alexandrus: Sorry to interrupt you, but tell me more about these weapons because something like that happened on our planet.

Chafa: Well, that weapon was built from this new mineral called Super Eurotanius. It took a long time to develop in liquid stage, so after we had it we threw it at Atlantis, and they had no chance. Their shields were no match for this weapon; it wiped out everything, including all life on the Red Land. We didn't agree to that, so the Zurdos were obsessed with this new weapon and they were thinking to destroy your home. They want to conquer the whole galaxy. They want to be the masters of the galaxy.

Alexandrus: Wait a minute, when you say conquer the whole galaxy, that means that we're not the only ones, am I right?

Chafa: You're right; we're not the only civilizations in this galaxy. The Zurdos knew that with a such weapon, they could invade your place, but they didn't realize that we didn't agree with them. The minute we found out, we sabotaged their plot, and we didn't give them time to develop it 100%.

Alexandrus: Excuse me, they sent their missiles to our planet, and they almost wiped us out.

Chafa: You're right in that, but the difference was that our warriors didn't allow full development from that attack. I would say that it was around 35% developed.

Alexandrus: Thirty-five percent. Wow, I don't want to imagine 100%.

Colonel Demello: Thirty-five percent what? Tell me, son.

Alexandrus: Sir, I think we should be thankful for these people.

Colonel Demello: Why, what did they do?

Alexandrus: They saved us, they didn't let the Zurdos develop the Super Eurotanius at its full strength.

Colonel Demello: What? You're trying to say that the Zurdos weren't trying to destroy us?

Alexandrus: No, sir, they were trying to wipe out everything, all life forms on our planet, but these people sabotaged their plan.

Colonel Demello: But why didn't they strike us back?

Atlacatl: Simple, they needed more time to develop the Super Eurotanius.

Colonel Demello: How much time?

Alexandrus: I don't know, sir. Chafa talked about it—it could be months, years, I'm not sure.

Atlacatl: We should act fast; we probably won't have enough time.

Colonel Demello: Son, ask Chapa.

Alexandrus: Chafa, sir.

Colonel Demello: Okay, whatever, ask him if they are going to join us.

Alexandrus: Sir, it's obvious that they will.

At that moment they were interrupted by someone approaching.

Chafa: Please make yourself comfortable. You are our new friends and you're in time for our ritual. We're offering a sacrifice to our god Chimbe.

Alexandrus: Sir, you're not going to like this; they're doing a sacrifice.

Colonel Demello: I can't believe it.

But that was just the beginning of the controversy. They were taken to a privileged place close to Chafa's throne. Some guys carried the wagon, and there were guards carrying swords waiting by the wagon, and then a woman walked out it. She was so beautiful that Alexandrus stood up right away.

Alexandrus: Sir, there's a woman.

Colonel Demello: Yeah, I see her.

Alexandrus: Do something, sir.

Colonel Demello: We can't interfere with their culture, but there must be a way to convince these people.

The guards were taking this young beautiful woman to the priest.

Professor Kishimoto: They're taking her to that priest. The Aztecs and Mayans used to do that. The priest was responsible for the sacrifices to their gods.

Alexandrus: Yeah, but that was a long time ago. The Spaniards brought Catholicism to America, and Jesus changed the human sacrifices for faith in one God, the savior of the world.

Atlacatl: We must do something.

Alexandrus: I agree.

The rebels were preparing for their ritual session, but the woman seemed not to agree with her sacrifice. She saw Alexandrus and asked for help as she passed by him.

Colonel Demello: What is she saying, son?

Alexandrus was silenced for a couple of seconds.

Alexandrus: This woman asked us for help and I'm going to help her.

Alexandrus moved toward her.

Alexandrus: Stop, you can't do this.

All the people around were petrified by Alexandrus's reaction; even Chafa was speechless. The guards lifted their arms against them quickly.

Alexandrus: Please let me tell you something with all due respect about your beliefs. I think you shouldn't do human sacrifices anymore, ever. Don't you see, people, every sacrifice that you're doing is wrong. We might need them in our fight with the Zurdos.

Chafa: Our god is going to punish us.

Alexandrus: God is not bad, He doesn't hate, He's love, comprehension, and respect. He doesn't like human sacrifices; we have to be together in this quest. If we are going to unite forces, we have to stop this right now. We don't do that back on our planet. Please, we're just asking for a moment of reflection. Besides, have you asked this woman to be sacrificed?

The people were murmuring among themselves.

Colonel Demello: Son, I hope that everything you're saying is convincing the people because otherwise there's going to be more sacrifices tonight, and I don't want to be a sacrifice for their gods. Get ready, everybody.

Everybody was getting ready for action.

Alexandrus: Please wait.

Chafa: Strangers from the Blue Land, do you have any idea what our customs are? Do you think that you can change our culture? Do you really think that we are going to allow that or accept your customs?

Alexandrus: No, not at all, we're not here to force our ideas on your society. We're here to destroy the Zurdos. On our planet, life is priceless, so if you think by killing people, your own people will be saved from the Zurdos, you're wrong. Let me ask you something— just ask this woman if she wants to be sacrificed.

Drucila broke her silence.

Drucila: (alien language) He's right, I don't want to die in vain.

Now everybody was shocked and confused.

Drucila: I've been telling Cacique Chafa that we don't need to do this. I told him that we need each individual of our tribes together; we can win this battle and we don't need to kill ourselves to please our gods. How many of our best warriors or our best people have been sacrificed to those gods in vain? They would have been useful right now.

Alexandrus: You see, this woman does not want to be sacrificed. Come on, people, this is the time to begin a new chapter in your life.

The older people were getting upset about what they were saying, and the young ones seemed to be in agreement. One of the young adults stood up.

Cayo: (alien language) We're not pleased about what our customs are; we're against human sacrifices, and we have been saying this long before you strangers showed up. You should've seen how many of our best warriors, scientists, the best of our people have been sacrificed and, worst of all, in vain, so we're with you strangers.

The rest of the young adults were chanting.

People: (alien language) Yeah, we're with the Blue Land, Blue Land.

Colonel Demello: Alexandrus, tell me what's going on, tell me that we're not going to be sacrificed.

Alexandrus: Sir, I don't think so, I think we've just revolutionized these people. The young ones are with us; it's just the older ones that are against us.

Chafa: You see, strangers, you can't just break our customs.

Drucila: Not them, but we can.

Chafa: You can't do this, we're going to be punished.

Drucila: Okay, how about you? (pointing at the priest, then some older men, and finally Chafa) or you Cacique Chafa, why not all of you?

Chafa: You know we can't die like that.

Drucila: Why not? Who gave you the right to choose who's going to live and who's going to die?

Chafa: You can't do this; we're going to die, all of us. Guards, take all the insurgents to the Hole.

The rebels spread out. Some of them didn't obey him, and of course some of them were loyal to Chafa.

Alexandrus: Okay, people, get ready because this thing is heating up.

Atlacatl: We're ready.

But all the people against Chafa were going to Alexandrus's side.

Cayo: We're with you, strangers from the Blue Land.

Drucila: Yeah, we pledge alliance to you strangers; we're glad that we're going to unite our troops.

Alexandrus: Well, you're very welcome.

Colonel Demello: I think we should leave immediately, guys.

He was holding his gun.

Alexandrus: Chafa, you have another chance. It's your people. Don't let this happen. We have something in common. Destroy the Zurdos.

Chafa: You strangers, you must leave this place, you have just broken our principles of life.

Alexandrus: I don't think so, Chafa, you're the only one responsible for this. It's your decision—you can stay here or leave with us. We're just looking for some friends and you didn't care about your people. They were against your will long before we even showed up, so we're leaving.

Colonel Demello: Alexandrus, how about these people?

He was looking at the new allies.

Alexandrus: They're with us, sir, they're coming.

Colonel Demello: How about the others?

Alexandrus: I don't know, sir, I think they will stay here.

Colonel Demello: But they're in danger here.

Alexandrus: We all are in danger, sir. We must be careful.

Atlacatl: I think everybody is ready to leave.

Colonel Demello: Okay, move out, people, move out.

Alexandrus: Come with us, Chafa, you don't have to do this.

Chafa: Never, we'll be here, and that's it.

Alexandrus: Okay, just watch your back, the Zurdos might come at any time.

Chafa: Just leave, you will regret this.

Alexandrus: All right.

Wasting no time, our heroes left the rebels' camp; they had new allies and they were just concentrating on Atonal's city. Chafa and his loyalists were very angry. As it turned out, they just started another big journey to save their lives.

Alexandrus (to Drucila): You must be a virgin gift for your gods.

Drucila: A kind of that, but thank you. My name is Drucila I'm a Doctorius in the higher ranks from Izalco, the second most important city in Alazan, and I also speak seven languages, including three old dialects.

Alexandrus: Oh, I'm impressed, now I see why you were chosen for the sacrifice, but I'm guessing that Alazan is the name of this place that we call Europa. You're also a doctor, I don't know your specialty yet, and you're a linguistic person. Seven languages, that's something special, and beautiful, wow.

Drucila: I have a question.

Alexandrus: Go, ahead, ask me.

Drucila: What do you mean by "virgin"?

Alexandrus (embarrassed): I'm…it means…I don't know…how do I explain it to you?

Drucila: Come on.

Alexandrus: Okay, it's when you and a man haven't been together sleeping, touching.

Drucila: Yeah, all women have to be in a ritual ceremony before they sleep with their man. There is respect, trust, and of course the most important of all, love.

Alexandrus: We call it marriage.

Drucila: Is it the same in your place?

Alexandrus: Yeah, right.

Colonel Demello came and saved Alexandrus from that tedious moment.

Colonel Demello: Anything good from these people?

Alexandrus: Yes sir, they are a great civilization. Drucila is almost perfect.

Colonel Demello: Watch out, son, I'm seeing love in the air.

Alexandrus: Not at all, sir, it's like, she's so beautiful, intelligent, with a good education, I'm just...

Colonel Demello: In love?

Alexandrus: No, I'm just fascinated. I would like to know her better. I'm just trying to be a good friend.

Colonel Demello: Well, son, my mom and my dad started just like that, and look at me now, I'm the fruit from that friendship. Okay, just be careful. Love is a strange thing to describe; when you are in it, everything seems wonderful with happiness and it's a great feeling, but when something is not going well, it may hurt you really bad.

Alexandrus: Not if you are really in love.

Colonel Demello: Okay, son, we should find a place to rest.

Alexandrus: Yes sir.

It was raining, and after a couple of hours, they settled at the top of a mountain to rest for a while. It took some time to set their camps up because of the heavy rain and winds, and they knew that the worst was coming. They were getting to know each other, sharing their cultures, customs, and they were also eating some food and some plants. The night was going to be very long. There were about ten rebels that joined them. The cold night wasn't much easier than the Zurdos, but it was making them sleep. Alexandrus was still awake. He was also injured from his past fights, and he was walking alone. He was watching the sky, but while the wind was blowing on his face and he was checking out his wounds, suddenly he heard some footsteps approaching. He grabbed his sword right away, and the next moment, he was astonished. Drucila came right to him, and there was a moment that they were looking at each other. They didn't talk after a couple of seconds; they were kissing each other. It was almost daylight. The rain had stopped, but there was another big electric storm coming. That electric storm was very dangerous, and the thunder sounded like a big blast back on Earth, but nothing could compare to it. In two long days of heavy rain, it killed one more soldier and a rebel when they were doing their job.

Dulcinea, Simbra, Lt. Dago, Amadeus and the rest of them had been struggling to survive the environmental conditions of the weather; the rain didn't stop at all for almost two days.

Amadeus: Oh my, we can go out. I never thought that we were going to be stuck in this place.

Dulcinea: Amadeus, at least we rested for a while; we needed this break. Thanks for the rain.

She was feeding Simbra.

Amadeus: All right, but I'm getting bored.

Lt. Dago: Hey guys, I need to talk to you. Captain J.J. is getting weird; his behavior is getting out of control.

Dulcinea: He was out of control before we landed. It was his fault. We shouldn't have crossed that river, ever, but it was his stupid pride that he couldn't handle.

Amadeus: Especially coming from a woman.

Dulcinea: We lost a lot of men, we almost got killed, and look what's happening. I think he's been drinking or taking some kind of drugs. The thing is, I don't care.

Lt. Dago: Hey, calm down, Dulcinea, we know you're right, but he's a human. He's depressed. I think he's learning from his mistakes.

Dulcinea: Yeah, but the people that we lost, they never got a second chance.

Amadeus: Relax, Dulcinea, all those people knew the risk of coming to Europa; they're casualties of war.

Dulcinea: How about their families, what are they going to tell them? They hope to see them back; they didn't deserve to die like that, especially knowing that we could have prevented it. We don't have enough people to fight, you know that.

Lt. Dago: I think you should go to see him. It will help him.

Amadeus: Okay, let's go.

Dulcinea: I'm not going. I've got to find a way to save this little girl from them.

Amadeus: Please, Dulcinea, do it for Simbra and us.

Dulcinea: All right. (alien language) I'll be back, Simbra, stay here.

They went to see Captain J.J., but Captain J.J. was so depressed. He had been using drugs that he brought from Earth. He was alone in the camp. He was hallucinating all the horrors that he saw in his life, and he saw the faces of destruction from what happened in Asia. He was so selfish, and then, the ghosts of horrors were haunting him again. He knew it was his mistake that almost got everybody killed. When they were crossing that river he knew it could have been prevented, but it was his pride, selfishness, and ineptitude that drove him to that point. There was no turning back. There were so few of them that in the next meeting against their enemy, they would have no chance even to give up. In his hallucinating, he saw the faces of the soldiers dying by the Zurdos and drowning; he also remembered Dulcinea's predictions. "We shouldn't cross the river, something is out there." He was a complete mess. He was hearing moans and more voices and then he started to run out of his camp. He had no idea where he was going, he was just running and running, trying to get away from those voices.

Captain J.J.: Please don't hurt me, it's my fault.

He started to run again. At that moment our heroes showed up in Captain J.J.'s camp.

Lt. Dago: (entering the tent) Sir, I brought Dulcinea and Amadeus.

Amadeus: (entering) Where's Captain J.J.?

Dulcinea: (from outside) Hey guys, he's not here.

Amadeus: (going outside) How do you know?

Dulcinea: Look at those injections, it looks like morphine.

Amadeus: He's been taking them.

Lt. Dago: (going outside) He's not there.

Dulcinea: He's out there, he should be somewhere.

Amadeus: I think he's in trouble, let's find him.

Lt. Dago: Okay, I'll go this way.

Amadeus: Okay, I'll go the other way.

Dulcinea: I'll stay here; I have to take care of Simbra.

Amadeus: Lt. Dago, be careful, it's raining a lot and we don't know this place.

Lt. Dago: Take care; I hope it's not too late.

Everybody left the camp.

Lt. Dago (shouting): Captain J.J., Captain, where are you, sir?

Amadeus was shouting too. He saw some footprints.

Amadeus (to himself): He's got to be here but where?

They continued looking for Captain J.J. but nothing. It had been two hours of searching, and he was very concerned about going so far. He could have been easy prey for his enemies; besides, he didn't know the place. Finally he got to the point of a rocky mountain, and he could see Captain J.J. on the top of a big rock.

Amadeus (to himself): I've got to hurry up.

But Captain J.J. was on the edge of a cliff and he was going to jump. He was still hearing those voices.

Captain J.J.: Please don't hurt me, don't.

Amadeus (running): Captain J.J., Captain J.J., don't do it!

Captain J.J. (crying): No, please, forgive me.

Amadeus: Sir, don't.

It was hard convincing him, but when Captain J.J.. was going to jump into the darkness of the cliff, Amadeus caught him by his arm just in time.

Amadeus: Sir, don't do it, we need you. There are some people that need your command. You did what you had to do, but this is the moment for leadership.

Captain J.J. was unconscious from the effect of the drugs.

Amadeus: It's all right, sir, we're going to the camp with the others.

They left the scary scenery. They were heading to the camp. Amadeus put him on his shoulders, sometimes stumbling hard on Europa's soil. In the camp was Lt. Dago and the rest of the group waiting for them.

Lt. Dago: Oh my God, you found him. Let's put him on his bed.

Amadeus: He's unconscious—let him rest and tomorrow he will be fine.

Lt. Dago: Thank you, Amadeus, you guys are great, not just warriors.

Amadeus: We're doing our best.

Captain J.J. was lying on his bed, and the rain was almost over. They had to decide the future of Captain J.J. They were wondering if the others guys were still alive, because they were hopeless and stranded in a place so far from their homes. After a couple of hours the rain stopped.

Dulcinea: Oh my God, the rain is over.

Simbra: (in English) Thank you for taking care of me.

Dulcinea: What? How do you know English?

Simbra: I have been listening to every word of your language.

Dulcinea: You are a smart girl; you never cease to amaze me.

Simbra: Thank you.

Lt. Dago (arriving): It's time to go. I think one of our scouts found something. There's a city.

Simbra: It must be Alazan. My parents brought me there.

Lt. Dago: Is she speaking English?

Dulcinea: Yeah, Lt. Dago, these people are very intelligent.

Lt. Dago: Okay, we should move right away.

Dulcinea: How about Captain J.J.?

Lt. Dago: Hey, he's much better now, he doesn't remember anything of what happened.

Dulcinea: Yeah, right. Let's go, Simbra, we should move.

Simbra: Okay.

Instead of regretting the past, they were heading to this new city that they just found, hoping that they could find help and food. But what they didn't know was that in Alazan, there was something else waiting…just for them.

Meanwhile Alexandrus's group were just proposing to move out.

Alexandrus: Good morning, Colonel Demello.

Colonel Demello: Good morning, son, I haven't seen you around lately.

Alexandrus: It's the weather, sir, but we've known that the next stop is Alazan; maybe we can find some help over there.

Colonel Demello: Are we getting close to Atonal's place?

Alexandrus: Well, Drucila said after crossing Alazan it's just a matter of time before Atonal finds us.

Colonel Demello: Good, so let's go to Alazan, and God bless us all.

Alexandrus: I'll tell the others.

This time they weren't ready for an imminent showdown against the Zurdos, but they were heading to the city of Alazan.

Dulcinea and Amadeus were also close to Alazan. They could see the city, but it wasn't what they were expecting.

Dulcinea: Wow, this city must have been beautiful before.

Amadeus: Yeah, what a waste.

The city seemed completely empty. Parts of its structures were completely damaged, and the streets were covered by dust and ashes. They could smell and feel the smog from the bombings. It seemed the Zurdos had just attacked Alazan.

Lt. Dago: But where is everybody?

Dulcinea: I don't know, but I don't like this.

Amadeus: Oh oh, I don't like to hear that, but you're right.

Lt. Dago: Do you think that everybody is dead?

Dulcinea: I'm not sure but it's strange—this city looks like there was somebody here, not long ago.

Amadeus: Days?

Dulcinea: No, hours.

Captain J.J. (approaching): What do you think, kids, anybody here?

Lt. Dago: Sir, it seems empty.

Dulcinea: What was that?

Dulcinea saw some shadows moving fast in a corner.

Amadeus: What?

Dulcinea: Over there, I'm not sure, what was that?

Amadeus: Sir, we should move and fast.

Captain J.J.: You're right, let's get out of here.

They started to move fast; then all of sudden they heard something in an alley.

Amadeus: Wait.

Captain J.J.: What happened?

Amadeus: Did you hear that?

Everybody was listening, looking around…nothing.

Captain J.J.: No, I don't hear anything.

Dulcinea: Shhhh, I can hear it too.

Lt. Dago: But what?

Amadeus pointed at an empty street.

Amadeus: It's coming from over there.

Dulcinea: (pointing in another direction) And there.

It was the sound of drums; the sound was getting closer and closer, and they didn't know what to do. They were surrounded by the beat of the drums.

Amadeus: It's coming from everywhere.

Dulcinea: There's no chance to escape.

She grabbed Simbra.

Lt. Dago: We're surrounded.

Suddenly, our heroes were trapped; they were just waiting for any attack, but they had no idea what was happening. Then the sound of the drums stopped.

Dulcinea: It stopped.

Amadeus: Look there, somebody's coming.

There were some shadows coming slowly from everywhere, and the sound started to hit again.

Lt. Dago: I don't like those shadows.

In a minute they were surrounded by those shadows, but there weren't any shadows at all, they were people, the ones supporting the Zurdos.

Dulcinea: They're people.

Captain J.J.: Are they rebels?

Simbra: No, they're really bad, they're Zurdos.

Those people looked very scary; some of them were wearing masks, some of them had their faces painted, some of them had swords, and some of them had strange animals kept on chains (they were something similar to a dog, but scarier). Our heroes were really in trouble.

Lt. Dago: Sir, we're outnumbered—should we attack?

Dulcinea: No, they are too many, we don't have a chance.

Amadeus: You're right; we don't have a chance at all.

Captain J.J.: Drop your weapons, we give up.

Lt. Dago: What? Are you serious, sir?

Captain J.J.: Do what I said.

Dulcinea: He's right, let's drop our weapons.

Instead of fighting them back, they dropped their weapons; after all, they didn't have any chance of destroying this new enemy. They were civilians affiliated with the Zurdos empire, and they escorted our heroes to their leader.

Amadeus: Where are they going to take us?

Dulcinea: No idea.

Lt. Dago: We shouldn't have given up; they're going to kill us anyway.

They were taken to a cell; they were prisoners of the Zurdos. Everybody was together, and the prison was really dark. They did not know what was around them much less how they were going to escape. It was a long night in that room.

Dulcinea: I don't see anything. Let's stay together, hold hands.

Amadeus: Why did they keep us alive?

Dulcinea: They have their own reasons.

Simbra: Dulcinea, I'm scared. I can't see anything.

Dulcinea: It's all right, just stay close to me.

Amadeus: Ssshhhh, something is coming.

Dulcinea: I'm sensing it too, there's a light coming.

Amadeus: Yeah, I'm hearing footsteps also.

There were some footsteps approaching. They couldn't see well, but there was a light coming from somewhere.

Dulcinea: Someone is coming.

There were some lights flashing in our heroes' faces, blinding them, and then a voice broke the silence.

Zurdos' voice: (alien language) Who's in charge?

Captain J.J.: This is it.

He stepped out of the group, wanting to attack.

Amadeus: Be careful, sir.

Zurdos warrior: (alien language) Why are you here? Who do you think you are to come to invade our place?

He kicked the captain's stomach.

Dulcinea (stepping out too): Hey, he doesn't speak your language.

Amadeus and Lt. Dago went to help Captain J.J.

Zurdos warrior: So, how do you know our language?

Dulcinea: Because I learned from my father. He used to live here; he was a scientist from Europa. We've come from planet Earth and we're not going to let you destroy our home; we're going to make you pay for that cowardly attack on us.

Zurdos warrior: Very impressive for a female in your place, but I think you don't get it. We'll take the Blue Land with or without your consent, and speaking of cowardly, your father and two others were the traitors from our mighty Empire. It would be a pleasure to destroy their daughter with my own hands.

Amadeus: Don't touch her, I will kill you with my own hands.

Zurdos warrior: Another one? Ja, ja, ja. It would be double the pleasure.

Angrily the Zurdos warrior grabbed his sword and was going to strike our heroes, but something stopped him; there was another voice.

Ateocoyo (alien language): That's enough. Take them to the Quezalte. They would love them.

Dulcinea: Quezalte?

Amadeus (to Dulcinea): I don't know what that is.

Simbra: Not there…no.

Dulcinea: Why? Who's Quezalte?

The Zurdos warriors left the prison, and once again it was completely dark. Simbra: It's a terrible place. My parents used to talk about it. No children were allowed to go there. It's a place where the rebels were punished.

Dulcinea: By who?

Simbra: I don't know, but my parents used to talk about it. All the citizens from this city were allowed to get in; it was a place where all the bad people were punished until they were dead.

Amadeus: Tell us more of that place.

Simbra: You're asking the wrong person, but my parents mentioned there were monsters inside.

Lt. Dago: Great, so there's no chance that we'll survive this.

Amadeus: Not so fast. We can still win this battle, we just need to focus on our enemy.

Dulcinea: And pray.

Lt. Dago: Sir, what should we do?

Captain J.J.: (pausing) Die with dignity.

Lt. Dago: What do you mean, sir?

Captain J.J.: This is it, wherever we're going, we're never going to make it out alive.

Dulcinea: I don't think so, sir. Your soldiers don't want to hear that.

Captain J.J.: Anything better?

Dulcinea: We were sent to this mission because we were the best on our planet, and we'll finish this mission because we will do it for our people.

Amadeus: She's right, we'll finish this journey because we're not going to let the Zurdos take out our home. Whoever makes it alive, they will finish it, all right?

Everybody: All right, all right!

Amadeus: Just rest as much as you can, because you will need it. And one more thing—it has been a pleasure to come all over from Earth to this place, Europa.

Dulcinea: Me too, guys, it's a pleasure.

Everybody: For Earth, yes, for Earth.

It had been a couple of hours since Ateocoyo was interrogating our heroes. There was a creaky noise in the dark; someone was opening the gate. It was already daylight, and at first the clarity of the daylight seemed to bother our heroes, but after a couple of seconds, they were okay.

Zurdos warrior: Move out, move out. The people are impatient.

Dulcinea: We need some water.

Zurdos warrior: You won't need it soon, believe me.

They were escorted to a kind of tunnel or corridor; it was halfway dark and halfway light.

Captain J.J.: Where are we going?

Amadeus: I suppose they're taking us to Quezalte.

Lt. Dago: But what exactly is Quezalte, some kind of warrior?

Captain J.J.: Or some kind of monster?

Amadeus: Or both, or something worse.

They were almost at the end of the corridor.

Amadeus: Ladies and gentlemen, whatever is out there, let's fight for Earth.

Everybody: For Earth, for Earth, for Earth.

Finally they got there, and there was a noisy sound of people, the sound of the Zurdos. They were chanting. Our heroes were in a kind of Roman coliseum. They were surrounded by a large wall, and the floor was dirt and sand. The crowd was getting impatient; they wanted to see some action. There were more warriors inside of the arena who were a kind of gladiator. Some of them were carrying pieces of human bodies. It looked like there were some people before them who simply didn't make it alive. The ground was covered with human blood, our heroes got scared, and the crowd was making a lot of noise.

Lt. Dago: What in God's name happened here?

Dulcinea: It's horrible, they're so cruel.

Simbra: (crying) Please let us go out, please, I don't want to die.

The other soldiers were scared but they were not alone.

Lt. Dago: Look, there's more people coming.

Dulcinea: They're rebels too.

But there was somebody else coming too; there was a Zurdos warrior in the group because he had failed a hard task.

Amadeus: There's a Zurdos warrior too.

Captain J.J.: Yeah, he's not going to make any difference.

There was a loud horn sound, followed by a heavy drumming beat. Someone showed up in the arena, dropping some weapons on the ground, but they weren't enough for all of them. The gladiators already had their weapons; they wanted to start the big show right away.

Host (alien language): Citizens of our big community of Alazan.

There was silence; the crowd was quiet.

Host: Our city has been invaded by aliens who want to destroy our mighty Atonal; these invaders are joining forces with the rebels.

Crowd: Ahhhgg.

Host: And worst of all, they're trying to steal our Super Eurotanius. These invaders from the Blue Land are coming here, to our sacred land, and you know what that means?

Crowd: (alien language) Quezalte, Quezalte, Quezalte!

Host (alien language): Yeah, you're right, Quezalte to the alien invaders and all of those rebels supporting them. Quezalte, citizens of Alazan, this is for all of you.

The host spoke to some guards who left right away.

Captain J.J.: Can someone tell me what's going on?

Amadeus: We're in an arena and those people want us dead.

Dulcinea: (to Simbra) Just stay behind me.

Simbra: (sniffing) I don't want to die here.

Dulcinea: No you won't, believe me.

Lt. Dago: Hey, over there, I think there are weapons.

Dulcinea: You're right, let's go.

But the rebels that were brought after them took them all.

Dulcinea (alien language): Hey, give us some?

Rebel #1 (alien language): I'm sorry, but you don't have any idea what's coming.

Dulcinea: Because of that, give us something to defend ourselves; we're defenseless.

And the rebels just ignored her.

Dulcinea: Oh my God, we don't have anything to defend ourselves with.

Amadeus: Okay, people. Just be extremely careful, stay behind Lt.

Dago, Dulcinea, and me, and let's see what's coming.

Our heroes were in big trouble. The rebels had taken all the swords that the Zurdos left on the ground.

Lt. Dago: These warriors took all of them.

Dulcinea: It's understandable; they're fighting for their lives.

The crowd was getting excited.

Dulcinea: Oh no, whatever is coming, they're there—look, the people are getting exited.

Amadeus: All right, we're ready.

Everybody inside of the arena were expecting the heroes' action; they were released, running wild. The crowd was really into it; those gladiators were killing machines.

Host: There, citizens of Alazan, is what you came for. Your fighters are here.

Ateocoyo was in a special place where he was being guarded by a high council of members and Zurdos warriors, all of them armed. The gladiators were ready to kill or be killed; they were there for Ateocoyo's entertainment.

Dulcinea: They're gladiators.

Amadeus: And you know what that means?

Dulcinea: Kill or be killed.

Amadeus: Yeah, all right, people, don't be scared. Just concentrate on their moves, and we might have a chance.

Ateocoyo (alien language): Let the festivities begin. Citizens of Alazan, this is for all of you.

Ateocoyo gave the signal, and the Zurdos warriors went right away to destroy everybody; they were merciless, strong, and worst of all, killers.

Lt. Dago: Okay, here they come.

Captain J.J.: Bring them on.

The Zurdos warriors were attacking first the rebels. It was a bloody battle, and they were facing some resistance. Some of the gladiators went after our heroes.

Amadeus: To your right, Lt. Dago, Dulcinea, go left, and I am going to take the two in the middle.

Amadeus was the first one to encounter the first two warriors. With a fast move (almost bending over his body) he grabbed one of the Zurdos's hands, then punched him in the belly with his knee. His rival didn't have a chance, so he dropped one of his swords and Amadeus took it immediately.

Amadeus: Now we're even.

The one attacking Dulcinea had a kind of dagger and was wearing a kind of Roman helmet. Dulcinea was just avoiding contact with those daggers. She was fast, and her skills were too much for this warrior; the gladiator couldn't follow any of her movement, so he decided to throw one of his daggers. (He had a lot.) Dulcinea's reaction was to pick up a piece of wood on the ground right away. The dagger was coming right at her, so she put the piece of wood close to her face to block the dagger. Dulcinea took it out fast, and she had a weapon to defend herself. There was another one coming towards Lt. Dago carrying a big spear.

Lt. Dago: Come on, get closer.

At that moment Captain J.J. approached.

Captain J.J.: You're not alone in this—together we have a better chance.

They were ready for their opponent.

The Zurdos warrior made the first attack, pulling hard with his spear. Lt. Dago and Captain J.J. luckily were alive from such a weapon.

Lt. Dago: Sir, the next one, you move on his back, and I'll push him back at you.

Captain J.J.: Okay.

They were spreading out. Their enemy was ready to counterattack. He lifted an axe that he had on his back and he struck back but missed. When he was going to lift his axe again from the ground where it was buried, Lt. Dago made a suicide move. He pushed with his body the warrior's chest and made him stumble; Captain J.J. pushed his back onto him, and the two warriors from Earth disarmed the warrior from Europa.

Lt. Dago: Sir, take that weapon, I'll take one of his swords.

Captain J.J.: Do it quickly.

Lt. Dago: Yes sir.

Now they had some weapons to defend themselves with. Captain J.J. was choking his adversary with his arm, and the warrior was trying to escape his grip.

Captain J.J.: Go on, Lieutenant, go help the others—take his weapons.

Lt. Dago: Yes sir.

Captain J.J. was merciless. It looked like the warrior was asking for forgiveness, but the captain didn't care. He seemed to enjoy that moment until he killed him and after that, Captain J.J. took the axe and he went to help the others. The arena was covered in blood from each side of the battlefield. Amadeus and Dulcinea were taking down their opponents, Lt. Dago and Captain J.J. were doing their parts, and the rebels also were succeeding in their combat. It looked like our heroes were winning. They were fighting so bravely that they got to the point of finishing them off.

Amadeus: I think we won.

Dulcinea: Yeah, but what's next.

The scenery was horrible. Our heroes of course suffered some casualties, the rebels too, but they were standing over the arena. The crowd was getting angry; they didn't like to see our heroes standing over their fallen heroes. There was a moment of silence until the host broke the silence.

Host: Citizens of Alazan, somehow, the intruders have defeated our first warriors.

Amadeus: What? What is he talking about? The first warriors?

Dulcinea: I don't know, just listen.

Host: The invaders have advanced to the next level and you know what that means?

Crowd (chanting): Quezalte, Quezalte, Quezalte!

Host: You guessed it, Quezalte.

Lt. Dago: What are they talking about?

Dulcinea: I don't know, but they're chanting Quezalte.

Captain J.J.: So it's not over yet, right?

Dulcinea: No sir, it's not over yet, it's just the beginning.

Crowd (chanting): Quezalte, Quezalte!

Host: Here you have them.

The crowd was really euphoric about this Quezalte. Our heroes didn't have any idea at all.

Amadeus: Now what?

Dulcinea: Just get ready, people.

At that moment the ground started shaking.

Amadeus: It's the ground or Quezalte.

Everybody was scared.

Dulcinea: No, over there.

She pointed at the ground a few meters from them. The ground was shaking, the crowd was excited, and our heroes were scared. There was something coming out from the ground, and there were three big cages rising that had a black curtain covering whatever was inside. The crowd was getting even more excited.

Dulcinea: Oh my God, what's inside of those cages?

Amadeus: We'll find out very soon.

Ateocoyo: (alien language) Release them.

The cages were trembling, and after Ateocoyo's order, the cages were opened.

Dulcinea: Oh my God.

Amadeus: This time we're in deep trouble.

Inside of each cage, there were their ultimate fighters—the "Q-fighters." They were Quezalte, a kind of hybrid warrior, and they were the ultimate killer gladiators. In one of the cages, a Quezalte warrior was carrying two big rustic sticks covered by sharp metals; the other one was carrying two axes. He had great skills with the weapons. The third one was covered by chains and spears. The minute the cages were opened, the first one to leave the cage was the one with the sticks, and he ran like crazy, his sticks wiggling around his body. Then the others were out of the cage too, and they were killing the rebels. The one carrying the axes had a strange move, and it was to follow his movements; his axes were almost spinning around him. But the last one seemed to be the toughest one. He was covered by chains, carrying spears that made him look very scary. They were out, and they were ready to fight.

Captain J.J. : What that heck is that?

Lt. Dago: I don't know, sir, but this time we don't have a chance.

Simbra: (alien language) Please don't hurt us.

The first warrior with the sticks took down one of three rebels standing in his way. It happened so fast and it was so horrible the way this rebel died; he was almost grinded by those sticks. The other one with the axes was so merciless that two of the rebel warriors didn't have a chance to escape; they were almost cut in half by those metal weapons, and when the rest of the rebels saw that it was impossible to defeat them, they started to run away, but they didn't realize that the third killer had so much strength and power. It was hard for the rebels to escape from the gladiators' spears, they couldn't get away, and in a few minutes they were finished off. All the rebels standing were killed.

Dulcinea: Did you see that? They're so powerful.

Amadeus: They're the perfect killing monsters.

Lt. Dago: Guys, what are we going to do?

Captain J.J.: This is it, we're done.

Amadeus: Not so fast, let's show them what we are made of.

The Quezalte warriors were on their way to attack our heroes, and the crowd was electrified. The one with the sticks took Dulcinea, Amadeus took the warrior with the axes, and the one with the spears, Lt. Dago and Captain J.J.. took him. Simbra was protected by three soldiers.

Dulcinea jumped; she flipped in the air and avoided the contact from the rustic sticks. Her enemy struck back, but she was so elastic that again she made her opponent miss again.

Amadeus was having a hard time. This monster was quick in his movements and with his axes. One second he was in front of Amadeus, the next he was behind him. Amadeus had to be smart and cautious— one mistake and he might lose this battle, which meant his life.

Lt. Dago and Captain J.J. were using some shields that they found from the gladiators lying on the ground. The third killer was slow in his movements, but he was more powerful than the others. He was practically just walking, throwing his spears.

Lt. Dago: Sir, here he comes again, another spear.

Captain J.J.: Just watch the spears and we might have a chance once he's out of spears.

Lt. Dago: I hope you're right. Let's run, sir.

Captain J.J.: Go to his left side and stay there.

Lt. Dago: It's suicide for you, sir.

Captain J.J.: Do as I said, Lieutenant.

Lt. Dago: Yes sir.

Dulcinea was also running away from those monsters' sticks.

Dulcinea: (to herself) I've got to find a way to defeat this guy, but how?

Amadeus was fighting with his sword against the axes; suddenly he had a chance to take one axe from his rival's hand.

Amadeus: Aha, let's see how you fight with just one axe.

Still this monster had the skill to move so fast. The crowd was excited, but they started to get impatient; they weren't ready to see their killer heroes having some difficulties against ours, and it was taking so long.

Inside of the arena was a complete battlefield. There were many bodies lying on the ground; some of them were completed, and some of them were not. The Quezalte warriors didn't think that it was going to take long but it was.

Our heroes made it clear that they were there for one reason: To destroy the Zurdos at any cost. Amadeus was the first one to get to his knees and he grabbed some sand and threw it into the enemy's eyes. Then he got up quickly and put his sword into his rival's neck.

Dulcinea, after struggling, made her comeback too. She had disarmed the Quezalte warrior, and they both also saw a sword lying on the ground. Whoever got the sword would win. They started to run. Back on Earth Dulcinea was trained in the 100-meter track so she took advantage of her skill and got there. It was a close race for the sword, but our heroine got it first and crushed his belly.

There was just one standing. The spear guy was completely out of spears.

Captain J.J.: Now he's out of spears.

Lt. Dago: Wait a minute, sir, he's covered by chains.

Captain J.J.: Come on, get closer.

The Quezalte warrior was spinning the chain above him; then he threw it at them a couple of times, and got Captain J.J. in his leg. He couldn't run.

Lt. Dago: Sir, are you all right?

Captain J.J. didn't move. When the Quezalte warrior was going to finish him off, he got stabbed through the chest by one of his spears.

Amadeus: Not so fast, today is your last day.

He threw the spear, and even though the warrior was stabbed through the chest, he still had some juice left to attack again. He got another hit from another direction; this time it was from Dulcinea.

Dulcinea: This one goes for all that you have killed.

That spear killed him right away. Our heroes had passed another big challenge, but not so fast—the crowd was wordless of course, and then all of sudden there was a marching sound around the top of the arena.

Dulcinea: Hey, what's going on up there?

Amadeus: I'm not sure, but it's probably nothing good.

Dulcinea: Come close.

She grabbed Simbra's shoulder.

Lt. Dago: Another surprise?

Dulcinea: Aha, here they come.

The marching sound was from the rest of the Zurdos troops; they were surrounding the whole area on the top and they were pointing their weapons right at our heroes, who at that point didn't have a chance against all those warriors.

Amadeus: How many are we in all?

Lt. Dago: We're eight standing and zero wounded.

Amadeus: Not a chance.

Captain J.J.: Let's give up.

Lt. Dago: Sir?

Captain J.J.: You don't see, we're defenseless.

Lt. Dago: But sir.

Captain J.J.: Don't question my order.

Amadeus: Sir, with all due respect, giving up is not the solution. They don't care.

Dulcinea: Yes sir, Amadeus is right, they don't want us alive. Look at what we've done, we have killed their heroes. What they want is revenge.

Captain J.J.: So what do you suggest, just do nothing?

Amadeus: Fight.

Lt. Dago: Yeah, let's fight.

Dulcinea: This is the only chance we have and God bless us all.

The Zurdos troops were ready, just waiting for Ateocoyo's order. He was in shock and he was upset also. He stood up from his throne and gave the order.

Ateocoyo: (alien language) Kill them all.

The crowd was euphoric but they weren't alone. From the crowd there was a big blast from a bomb. It was Alexandrus and his pals, disguised as Alazan citizens. They had their own weapons and they started another big battle. But in that time Ateocoyo had jumped into the arena and he had beaten so bad Amadeus, Captain J.J., and Lt. Dago.

Alexandrus: Not so fast, go, get them!

Meanwhile our heroes were battling against the Zurdos in the stands. Dulcinea and the rest of her group were fighting also, and they were not given any break at all. To begin with, Alexandrus saw that Ateocoyo was coming at them.

Alexandrus: Hey Atlacatl, take care of these two.

Atlacatl: Where are you going?

Alexandrus: I'm not going anywhere; I'm just waiting for him.

He pointed in Ateocoyo's direction.

Atlacatl: He's their leader, right?

Alexandrus: Yes, he's the leader.

Atlacatl: Be careful,

Alexandrus: You too,

Atlacatl continued fighting with the Zurdos troops, and Alexandrus was just waiting for Ateocoyo's approach. Ateocoyo was carrying an axe with two blades on each end. He was furious and he was killing on his way too. He got to where Alexandrus was waiting.

Ateocoyo (alien language): Die, intruders.

Alexandrus (alien language): We're not intruders, we're just protecting Earth.

Both warriors were clashing; both of them were great. The ground was covered with people lying on the ground, blood, and everyone was fighting for their lives, but there was no escape; there was no safe spot to hide. It was just a battle for their lives. But our heroes were taking control of the situation. Dulcinea and three soldiers were fighting inside of the arena, and one soldier was protecting Simbra.

Alexandrus showed his skills in his match. He had to be very cautious, especially when Ateocoyo's axe was spinning around him, cutting everything in its way. It was made from Super Eurotanius.

The battle of Alazan seemed to last forever. It started to rain, but that didn't bother anybody. The crowd was running away and Ateocoyo was killing everybody on his way, including the people of Alazan.

Colonel Demello: Oh my God, there are a lot of warriors; we're never going to finish them off!

Atlacatl: Let's just finish this, we're almost done.

Certainly, our heroes were winning this horrendous battle, but the match that seemed unfair resulted in a big surprise. Alexandrus was getting rid of Ateocoyo. First Ateocoyo missed him in a counterattack with his axe. Alexandrus, with his powerful legs kicked his belly and then put an end to him with his sword.

Alexandrus: This ends here, no more killings.

He finished him off. Our heroes had found a way to overcome the Zurdos one more time. They were tired, bleeding, and exhausted, but they were ready for the final showdown against Atonal and his ferocious warriors.

Alexandrus: We have won once again.

Colonel Demello: Yes, son, we've won. We're unstoppable now; victory is just a matter of time.

Alexandrus: I wouldn't say that.

Colonel Demello: What are you talking about? You don't see we've won?

Alexandrus: What we did today is just the first step—the worst is still to come. They know that we're coming and they're ready for us.

Colonel Demello: I thought this was the city.

Alexandrus: This is Alazan; the City of Fury is the next stop.

Colonel Demello: Okay, and I was thinking of taking some souvenirs.

Dulcinea and Amadeus were coming toward them.

Dulcinea: Thank God you're alive.

She hugged him.

Alexandrus: Well, I'm happy to see you too.

Amadeus: Hey, thank you, guys. I would call this perfect timing.

He was bleeding from his nose and holding one arm.

Alexandrus: Don't mention it, you'd have done the same thing.

Colonel Demello: I'm happy to see you alive, kids.

Dulcinea: Thank you, sir.

Colonel Demello: Where's Captain J.J.?

Amadeus: He's wounded, but he will survive.

Dulcinea: Wow, you don't know what you've missed.

Alexandrus: I can imagine it.

Colonel Demello: The same to you guys; we've been through a lot.

Alexandrus: I'm glad that you've come so far. Together we'll destroy the City of Fury. I know that you guys have been through a lot.

Colonel Demello: Where are the others?

Dulcinea: They didn't make it.

Amadeus: How about the others, sir?

Colonel Demello: There are no others; we're just by ourselves.

Professor Kishimoto and Drucila arrived with a group of rebels.

Drucila (alien language): I'm glad you're alive.

She hugged Alexandrus.

Dulcinea (jealously): Who's that girl?

Alexandrus: Well, this is Drucila. She belongs to Europa's rebels. They're our allies.

Dulcinea: Well, you look like you've been in good company.

Alexandrus: They're just our friends; she healed me with some miracle plants.

Colonel Demello: (speaking softly) Somebody is jealous.

Dulcinea: No sir, I'm not jealous.

Colonel Demello: How do you know?

Dulcinea: Remember why we were brought here?

Colonel Demello: Oh, I forgot that you could read minds.

Dulcinea: And you too, Alexandrus. I'm not jealous.

She turned her back to them.

Professor Kishimoto: Oh my!

He was filming with his video camera.

Lt. Dago: Sir, I'm glad to see you.

He was walking with some difficulty.

Colonel Demello: Me too. Where's Captain J.J.?

Lt. Dago: He's coming—over there.

Colonel Demello: How many are standing?

Lt. Dago: Well, sir, we're eight, including a little girl and you guys. Let me see...

He started to count one by one.

Alexandrus: We have twenty-five people, and we have to finish this.

Dulcinea: Yes, we will do it.

Colonel Demello: I like your attitudes, kids, but we're going to need more than twenty-five people, not counting this little girl and the five we have wounded. We need at least one thousand for this quest.

Alexandrus: You're right but you don't see we've come so far. What we need is to stick together and believe in ourselves. People on Earth are hoping that we finish our mission, and these people from Europa are counting on us too, so let's finish it, guys. I know you've been through a lot, but just a little more...please.

Amadeus: I'm with you, Alexandrus.

Dulcinea: Me too.

Atlacatl: You know me, I'm in.

Colonel Demello: Let's rock and roll.

Captain J.J. was carried by one soldier, and Simbra was with them too.

Captain J.J.: Sir, I'm reporting that we failed you.

Colonel Demello: No Captain, you haven't failed me at all; you guys have made me proud. In my whole career I've never felt so proud to be in a group of people like you. You're my family here, and like Alexandrus said, let's do it for our people on Earth.

Alexandrus: For people on Earth.

Everybody: Yeah, for people on Earth.

Alexandrus: For the people on Europa.

Everybody: For the people on Europa.

They were motivated by Alexandrus' speech, but they were very tired.

Colonel Demello: Okay, listen up, we will rest for a while and then we're heading up to the City of Fury.

Everybody: Yeah.

They were taking care of the wounded ones, eating and joking. They also made small groups in the "restoring" area. In one of the groups were Alexandrus, Dulcinea, Drucila, and Simbra.

Dulcinea (alien language): So, do you think that we can recruit more of your people?

Drucila (alien language): I don't think so. Our people have been hidden, and most of our communities have been wiped out.

Alexandrus: (alien language) But there should be something, something that the Zurdos might be scared of.

Dulcinea: Since we've come to this place I haven't seen this super mineral that we called Super Eurotanius.

Alexandrus: You got it.

Dulcinea: What, what did I say?

Alexandrus: (alien language) Hey, Simbra, is there a Super Eurotanius mine around here?

Simbra: Not that I know of.

Drucila: What about the Valley of silence? They have that mineral.

Dulcinea: Where's that?

Alexandrus: What? I don't like the name of it, but let's go.

Drucila: You don't have any idea what's going on over there, it's a suicide; everyone who tried to cross the Valley of silence, no one has survived to tell the story.

Alexandrus: But we need this mineral, we have to destroy Atonal, his army, and we have to figure out our way back home.

Dulcinea: Don't tell me that you're going to say for oil.

Alexandrus: Well yeah, but no, I mean, we need combustibles for our way back home.

Dulcinea: You're right, we lost it in the crash.

Drucila: What's going on?

Alexandrus: Nothing important. You were saying about this Valley of silence.

Drucila: It's been taken over by one of the bloodiest warriors in Atonal's power.

Alexandrus: Well, that is new.

Drucila: But this one is so different from all the other ones.

Dulcinea: What makes him so special?

Drucila: You can't kill him.

Alexandrus: I've heard that story before.

Drucila: You don't know what he's capable of.

Alexandrus: Coming here, it was a tough mission, and finishing it is worse. It would be a miracle to get so far, but we have to go there to destroy him and his army. If we fail, nothing is going to stop them.

Drucila: He doesn't need any army—he's alone.

Alexandrus: Well, it's much easier that way.

Drucila: He has destroyed entire platoons that we have sent before.

Dulcinea: But we don't have any choice, we have to go there; it's the only chance that we can destroy Atonal and go home.

Amadeus and Atlacatl arrived with pieces of wood to make a fire. The night was cold, and it looked like it was going to rain…again.

Amadeus: Hey guys, we're bringing some wood. This will keep us warm.

Alexandrus: Thanks a lot.

Atlacatl: It looks like a big storm is coming up.

Dulcinea: Well, we better get ready. I'm freezing.

Alexandrus: She's right, let's get some rest. We'll need it.

Dulcinea: See you later, guys.

Atlacatl: I'm sleeping.

Amadeus: Okay, let's go.

Alexandrus and Drucila were alone.

Drucila (alien language): Hey, thank you.

She hugged him.

Alexandrus: For what?

Drucila: For what you did today. That was so brave.

Alexandrus: I didn't do it alone, we did it.

Drucila: That was a big victory for our cause.

Alexandrus: Just remember one thing: there will come better days.

Drucila: I hope so.

Alexandrus: Well, after this storm is over, we'll finish our mission. I don't know how, but we have to survive.

Drucila: I know, just take care of yourself. I'll see you later.

Alexandrus: You too.

Drucila took Alexandrus' head with her hands, and then she nodded her nose with his.

Drucila: I don't know what's happening to me, but you make me feel safe.

Alexandrus: Don't worry, I'll protect you.

Colonel Demello came out and faked that he was coughing.

Colonel Demello: Sorry to bother you.

Alexandrus: Don't worry, sir, we're just leaving. See you later.

Alexandrus and Drucila left in different directions, and Colonel Demello stayed alone. He was watching the sky, the stars, and looking at some pictures of his family with a lamp. He was looking at the rest of the majestic landscape of Europa and smoking a Cuban cigarette; he stayed the whole night alone, just him and the rain, then he fell asleep.

It had been a couple of hours after the rain had stopped, and everybody got ready.

Colonel Demello: Good morning, everybody, I hope everybody slept well, because if you didn't, don't worry, you won't need it.

Everybody: Yeah.

Colonel Demello: So, let's finish this journey and make our families proud.

Everybody: Yeah.

They were ready. They were going to the Valley of silence, and their spirits were high. They were singing, talking, and some of them were worried.

They had walked almost the whole day to the Valley of silence. The first group were Colonel Demello, Alexandrus, Atlacatl, Amadeus, Lt. Dago, and Captain J.J.

Captain J.J.: Now what? I think we're going in circles.

Colonel Demello: What do you mean, Captain?

Captain J.J.: We're lost, that woman is just toying with us.

Alexandrus: Sir, I don't think that we're lost. You don't see, it's too dark. We're good on this road. One more thing, I don't think she's toying with us, she's helping us, sir.

Colonel Demello: That's enough for me. We'll wait here, we need some rest. Let's unpack and tomorrow we'll have a better chance to see in daylight.

Lt. Dago: What's that, over there?

Amadeus: There are some lights.

There were four bright lights in the darkness of the Valley. Those lights were far from our heroes, for the moment…

Lt. Dago: It could be a village?

Colonel Demello: Or the Zurdos?

Alexandrus: It could be both, but there's something strange in those lights.

Colonel Demello: What, son?

Alexandrus: It's definitely not a village.

Amadeus: You're right, there's something that I don't like about those lights.

Colonel Demello: Probably there are some rebels lost in this valley.

Alexandrus: It could be. There's only one way to find out. Let's go.

Amadeus: Yeah, let's go.

Colonel Demello: Lt. Dago, tell the others where we're going.

Lt. Dago: Yes sir.

They left to find those strange lights, but to their surprise the lights were right on them too.

Alexandrus: Sir, I think the lights are coming at us.

Amadeus: They're moving.

Colonel Demello: What should we do?

Alexandrus: Let's just wait here, it could be anything.

Captain J.J.: Sir, we might be easy prey here. I suggest that we should go to those lights.

Colonel Demello: We'll stay here.

The rest of the others were coming too.

Dulcinea: What's happening?

Alexandrus: We're waiting here, it seems that those lights are coming to us.

Dulcinea: What?

Amadeus: Yeah, they're coming fast.

Dulcinea: Any way to hide?

Alexandrus: Too late, I think they spotted us.

Dulcinea: Zurdos troops?

Alexandrus: I don't know.

Dulcinea: I don't like this, let's get ready.

Lt. Dago: Oh my God, they're here.

There were four bright balls a couple of feet away from our heroes; they were in a kind of rhythm, one following the other one, one after another.

Alexandrus: They're getting closer.

Colonel Demello: Listen, everybody, watch out!

Drucila got closer to Alexandrus and grabbed his arm.

Drucila: Please watch your step; I don't want you to get hurt.

Alexandrus: Don't worry, I'll protect you.

Drucila: You don't have any idea what's coming?

Alexandrus: We'll know soon.

Drucila: No matter what happens, you will always be in my heart.

Alexandrus): What do you mean?

Drucila's comments surprised Alexandrus, but it wasn't important at that time; the mysterious light balls were there.

Captain J.J.: What's that?

Amadeus: No idea, but be careful.

The lights didn't move; they were just watching our heroes.

Beyond those lights, there was a human shape; it was really dark and they couldn't tell who was behind it. The shadow was getting closer.

Alexandrus: Somebody is coming.

Colonel Demello: Hold your position, somebody is coming.

Dulcinea: Who's that?

Amadeus: It seems like…a man.

It seemed like a human dressed in black; he was completely dark, and his hands have some strange power over those lights. He had total control of them.

Amadeus: Why did he stop?

Alexandrus: I don't know, it looks like he wants to ask us something.

Amadeus: Like what?

The dark figure broke the ice. He said something in an ancient alien language.

Colonel Demello: What is he talking about?

Dulcinea: I don't have any idea what language he is speaking.

Alexandrus: And you, Amadeus?

Amadeus: Not me.

Dulcinea: My parents didn't teach me this language; I would've known it right away. It seems to be some ancient dialect.

Alexandrus: Okay, I'll go first to find out.

Drucila stepped up right away.

Drucila: Don't go.

Alexandrus: What?

Drucila started to say something in the ancient dialect to the Dark Knight.

Alexandrus: What did you tell him?

Dulcinea: Whatever she told him, it got him angry.

Amadeus: Yeah, she's right, we better leave.

All of sudden the lights started to move again, but this time our heroes had a better look at those flying balls; they were circular saw blades, and they were heading toward them.

Alexandrus: Watch out, those are saw blades!

The circular saw blades were really fast; in the first assault they cut two rebels warrior into pieces in just seconds.

Dulcinea: Oh my God, what did she tell this guy?

Amadeus: It's too late, here they come again.

The second assault also had a devastating effect; it took two more warriors down.

Colonel Demello: Run, save your lives.

Lt. Dago: Sir, over here.

Lt. Dago took Colonel Demello and Simbra into some kind of a hole in the ground.

Colonel Demello: I think we're not going to get away from this one.

Lt. Dago: God bless us all.

Simbra was crying. Colonel Demello, Lt. Dago, and Captain J.J. were freaked.

Not far from them, Amadeus was hidden behind a big tree.

Amadeus (telepathically): Are they coming?

Dulcinea: I don't know, where are they?

Amadeus: I think we're lost.

Alexandrus was also lost. He was trying to find the Dark Knight, thinking maybe he had a chance to defeat him first and not to deal with the blades.

Alexandrus (to himself): Hey guys, are you there? Where are you?

Nobody was responding, and then he heard the sound from the

EFRAIN CORDERO

circular saw blades coming at him.

Alexandrus: Oh, no.

He had a chance to run away, but he had to face them. The first saw blade struck but he avoided it, and it crashed into a big rock; then the second and third were coming. With his sword he put them out of reach from the Dark Knight. He swung his sword like a baseball bat and hit them hard, but not before losing his sword with such impact. The fourth and final saw blade was without movement, but the Dark Knight was coming too. Alexandrus was ready for anything. Suddenly the fourth saw blade started to move and the Dark Knight attacked him.

Alexandrus: Aha, two against one, that's not fair.

Another sudden death combat started again, with Alexandrus fighting for his life, and the other unknown warrior protecting something. They were extraordinarily good, and their movements were amazing. Alexandrus also had to fight with the saw blade. He had to concentrate on both threats. He started to run away in the forest; he was unarmed and the blade was after him. The harder he ran, the closer the blade got to its target. He stumbled on a small rock that he didn't see, and the blade crashed against a rock. Luckily he was saved but not before it got its target. It rubbed inches away from his arm.

Alexandrus: Damn, it almost got me.

The Dark Knight was coming. Then again they started to combat until the end there were no breaks, there was nothing. Alexandrus grabbed a piece of wood that was lying on the ground. He was trying to defend himself, but he was very tired and bleeding. All of a sudden the circular saw blade stared to fly again from the rock. Alexandrus had to show why he was a good leader—with an acrobatic move, he made the blade follow him, and when it was time he made the blade crash into the Dark Knight's shoulders. He was out of options.

Alexandrus: I don't know what language you are speaking, but I need to know. Why didn't you attack when you had the chance? We're from Earth and we want to destroy the Zurdos. What did you tell us first?

He was stumbling and fainting. The Dark Knight was wounded too, but he was trying to talk in the ancient language.

Alexandrus: What?

The Dark Knight barely could talk. Defeated by his own saw blade, he was losing a lot of blood. Then suddenly, he made some gestures to get Alexandrus closer.

Alexandrus: What? What do you want me to do?

He came closer to the wounded one, and the Dark Knight whispered something in Alexandrus's ear. He was in shock. What he was hearing from his adversary was very important. Then a ferocious spear coming from behind Alexandrus put the Dark Knight to his end. Alexandrus slowly turned his head where the spear came from and to his surprise, it was Drucila.

Alexandrus: You.

Drucila started to laugh out loud.

Drucila: Fool, you don't see, you did us a favor to get rid of the Dark Knight.

She started to change her appearance. Her skin was peeling off, she took off her clothes, and she was completely made of metal. She had two faces, one in front and the other one in the back. She had two sharp swords in each hand.

Drucila: Now, die.

She attacked Alexandrus, who was still shocked by what he was witnessing.

Alexandrus: Why?

From nowhere a spear targeted Drucila. She turned around and Amadeus was there. She started to attack Amadeus, but Dulcinea threw another spear at her. This one did more serious damage; it stabbed one of her arms.

Dulcinea: Take this, b..ch.

Drucila had no chance against all three, so she started to get away.

Alexandrus: Let her go, I know what we need.

Dulcinea: Are you all right?

Alexandrus: I can't believe I didn't realize that we were hanging around with the enemy.

Dulcinea: Don't worry, even us believed her.

Everybody was arriving.

Captain J.J.: Not me. Since the beginning I knew there was something wrong with this woman or whatever she is.

Colonel Demello: Are you okay, son?

Alexandrus: Yes, I'm okay.

Dulcinea: Hey, you're bleeding, let me help you.

Alexandrus: I will survive, don't worry.

Dulcinea: You were blinded by this monster.

Alexandrus: But there's something wrong with her.

Dulcinea: We know that already.

Alexandrus: No, I mean, she has two personalities at the same time. I think one of them was speaking how she felt.

Dulcinea: Come on, boy, she's part of the Zurdos.

Alexandrus: I know that, but...

Colonel Demello: I'm sorry to interrupt your lovely relationship, guys, but did this guy tell you something before he died?

Alexandrus: Yes, he told me where he was hiding the Super Euro-tanius mineral. He could have been a good ally, too. In others words unknowingly we destroyed our enemy's nemesis.

Colonel Demello: No, son, we did a good thing; with this mineral, we can destroy our enemy.

Amadeus: Well, I think we should keep going before Drucila gets the news to Atonal.

Colonel Demello: You heard him, let's keep going. Hmm…but where, son?

Alexandrus: Follow me. If I'm not wrong, the place should be there.

He pointed at a rock on the top of the mountain.

Lt. Dago and Atlacatl arrived.

Lt. Dago: Sir, we have four people down.

Atlacatl: Actually five, sir, counting Drucila.

Colonel Demello: So that makes us twenty-one.

Lt. Dago: Yes sir.

Colonel Demello: Well, let's go, people.

They began climbing the mountain in search of the mysterious rock that was keeping the mineral safe. It was already daylight, but there was fog, and once again it looked like a big storm was coming.

Alexandrus: We better get there fast, there's a storm coming again.

Dulcinea: We're almost there.

Alexandrus: Hey, Dulcinea, sorry.

Dulcinea: Sorry for what?

Alexandrus: For the inconvenience with Drucila.

Dulcinea: Forget it.

Alexandrus: But I felt so bad.

Dulcinea: I don't want to talk about it please.

Alexandrus: All right.

They were climbing the difficult road on the hill and helping each other. Alexandrus was in no shape for walking but he was leaning on Atlacatl's shoulders. They were very devastated about Drucila's personalities, especially Alexandrus that he trusted her too much. Finally they got to the top of the hill.

Dulcinea: Well, that's the rock. Now what?

Alexandrus (looking around): It should be here.

Dulcinea: Are you sure? Because I don't see it.

Alexandrus: Just be patient.

Atlacatl: Lt. Dago, what exactly is the shape of this mineral?

Lt. Dago: I have no idea.

Atlacatl: I hope we find it soon. If the Zurdos show up, we're easy prey.

Lt. Dago: I know.

Alexandrus: I got it! I've found the Super Eurotanius.

Everybody: Yeah.

Some of them were whistling.

Dulcinea: Wow, I never thought I'd find this mineral in that stage.

It was a cubic tube the size of a human arm, rainbow-colored, and it looked like crystal, but there was only one.

Colonel Demello: Just one?

Alexandrus: This is enough to destroy an entire city.

Colonel Demello: What about the rest of it? We need to take some to Earth.

Alexandrus: Sorry, sir, but this is the only one that the Dark Knight mentioned.

Dulcinea: I think it's the best way for us, sir. We don't want to end up like this planet.

Colonel Demello: Okay, let me have it.

Alexandrus: Here, sir.

Colonel Demello: Wow, fantastic, it's beautiful. You can feel the greatness in your hands.

Lt. Dago: It could be the radiation, sir.

Colonel Demello: Quiet, Lieutenant.

Alexandrus: It's okay, sir, there's no radiation in that.

Captain J.J.: Now what?

Alexandrus: We're going to Fury, ladies and gentlemen, nonstop.

Colonel Demello: Yeah.

Amadeus: People, you better watch this, you will be shocked! Look, there!

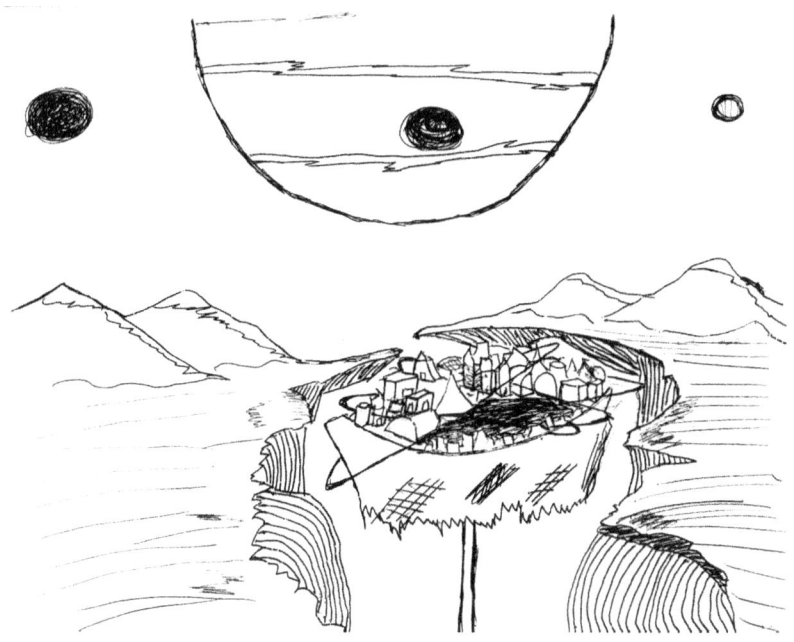

He was pointing to the city of Fury. It was a marvelous, panoramic scene, and our heroes couldn't believe that they were witnessing the enemy's fortress.

Alexandrus: Well, Atonal, here we go.

Our heroes finally got to their destination, the city of Fury. It had beautiful architecture, and the city was supported by a thin but strong column at the bottom. It was surrounded by the majestic mountains of Europa. It was a powerful civilization that once ruled the whole of Europa, but somehow it fell into the wrong hands.

Alexandrus: Amazing.

Dulcinea: It's beautiful, there's no place like it.

Colonel Demello: What a big civilization.

Amadeus: It's a pity that the Zurdos are taking control of it.

Dulcinea: Hey, Alexandrus, I've got a question. I still can't figure out how you knew exactly the place where the Dark Knight was talking about, speaking only God knows what language.

Alexandrus: Don't be surprised, but he told me in English.

Dulcinea: What?

Alexandrus: Before he died he told me in the English language; it was an ancient dialect they used to speak a long time ago.

Dulcinea: Why did he attack us?

Alexandrus: He thought that we were allied with the Zurdos.

Colonel Demello: Okay, people, we have the Super Eurotanius. Now let's make them pay for what they did to our planet.

Amadeus: Sir, but it's just one shot.

Colonel Demello: I know, son, but this is it.

Dulcinea: Sir, they might have innocent people living out there.

Alexandrus: She's right, there might be civilians down there.

Lt. Dago: How do you know?

Alexandrus: Well, we should send a patrol over there.

Atlacatl: I'm in, Lt. Dago: Count on me too.

Alexandrus: Let's attack first. Just remember the phrase "Whoever attacks first, attacks twice."

Lt. Dago: We'll be there in a couple of hours; the night will be long.

Colonel Demello: We'll wait for night and then we just throw the mineral and boom, they're out, we're in, and that's it.

Alexandrus: Sounds easy, sir, but you're the boss.

Wasting no time Lt. Dago and Atlacatl headed to the City of Fury to verify if there were civilians living in the city before launching the attack. They were waiting for nightfall. Not one of them had moved from their positions; they didn't want to make them suspect that they were there.

Dulcinea brought a hot beverage to Alexandrus, who had a high fever from his wounds. He was a complete mess from the earlier combat, and he wasn't ready for the final assault.

Dulcinea: Hey, do you want to drink some hot tea?

Alexandrus: Tea?

Dulcinea: Well, Simbra showed me some plants that her parents used to drink. It's like a tea but it heals everything. You need it right away; you have lost a lot of blood from the fighting.

Alexandrus: Thanks a lot. You can guide the rest of the platoon; I'm not sure if I'm going to make it from this.

Dulcinea: Don't talk, just rest. You will feel much better. We need you.

Alexandrus didn't talk anymore; he just fell asleep from the tea that Dulcinea brought. Night felt down, the rain started again, and the hours were passing by. Lt. Dago and Atlacatl showed up very tired from the mission.

Colonel Demello: It's about time, boys—what's happening down there?

Lt. Dago: There are no civilians in the city, just Zurdos troops, sir.

Colonel Demello: Perfect. Let's do it right now.

They brought some kind of device especially for delivering rockets a certain distance.

Colonel Demello: Ready?

Lt. Dago: Ready?

Colonel Demello: God bless the Earth now.

The projectile was launched, carrying the Super Eurotanius. It was headed to Atonal's city. They knew that it could be their only chance to destroy Atonal and his troops. Our heroes were outnumbered anyway, so they were just waiting for the impact and then the miracle.

The projectile was getting close, closer, and finally got to the target. There was a big blast, with a blue and yellow color surrounding the city. It looked like there were no survivors of such an explosion, but the fog and the smoke made it visibility very difficult for the moment. Our heroes were starting to celebrate their victory.

Colonel Demello: Ladies and gentlemen, Atonal is destroyed.

Amadeus: I would like to think that, sir.

Colonel Demello: Son, don't you see, we're victorious.

Dulcinea: We have to wait a couple of hours for the radiation and then we can search for survivors.

Captain J.J.: There are no survivors, lady, I can assure you of that.

Dulcinea: Let's see.

Colonel Demello: We'll wait for the daylight and then we'll go.

Everybody: Yeah.

Amadeus: We better be careful. I wouldn't like the idea going without Alexandrus.

Dulcinea: I know, he makes us feel much better.

Atlacatl: Be optimistic, guys, cheer up.

Amadeus: You're right, let's see what happens.

Dulcinea: Let's take a rest while we can. I'll go check out Alexandrus, see how he's doing. We need him so badly for the final assault.

Amadeus: Yeah.

The city looked very desolate. There was nobody standing up. The blast was so destructive that it almost wiped out the whole area. The rain was still falling, lightning was flashing, and thunder was roaring the whole night. Our people were settled without choice. Dulcinea stayed for so long, waiting to see if Alexandrus would get over from his fever. She took good care of him. Atlacatl was with three soldiers of his platoon, giving the final moments of the journey. They were sharing the moments that they were fighting together back on Earth. Amadeus and Lt. Dago were playing cards with two more soldiers, and the few rebels left were dreaming that some day they would live in peace once again. Colonel Demello and Captain J.J. were very satisfied from the attack but they were still wondering if anyone was still alive.

The sun was setting on the horizon of Europa, and our heroes were getting ready. Some of them were watching the city that once ruled Europa; others were giving the last touch to their weapons, and a few were writing letters to their families just in case they didn't make it.

Amadeus: So this is it.

He stood up from the ground.

Atlacatl: So be it.

Colonel Demello: Let's finish them once for all.

Meanwhile Alexandrus was waking up, and the first thing that he saw was Dulcinea lying on the ground by his side. He knew that she had taken good care of him, and he was feeling much better thanks to the medicine he drank.

Dulcinea (waking up): How's your wound?

Alexandrus: Much better. Hey, thanks a lot, you took very good care of me. I'm very pleased. I don't know how to repay you

Dulcinea: Just guide us to the final assault. We need you.

Alexandrus: I'm not sure if I can guide you guys; just look at what it happened back in the valley with Drucila.

Dulcinea: What happened back there wasn't your fault, she tricked us.

Alexandrus: Sorry, but it was my fault. Everything was a setup, and we destroyed the only ally that we might find in the whole Europa. He could have been very useful to us and now what? We're heading to a suicide mission.

Dulcinea: Well, a lot things happened while you were unconscious. We already launched the attack on the City of Fury, and apparently it seems that we destroyed all of them, but…

Alexandrus: You're not sure of that, right?

Dulcinea: Yeah, I'm not sure of that.

Alexandrus: I hope that they're gone; if not we're done.

Dulcinea: Please come with us.

Alexandrus: I can't do it anymore. I don't want to take all of you to a trap again.

He stood up and he was going to walk away when Dulcinea grabbed one of his arms.

Dulcinea: Please, we need you and I need you.

Alexandrus: My god, you're so beautiful.

They were looking at each other, but they were interrupted by Colonel Demello, who stepped out from the group.

Colonel Demello: I'm glad to see you alive, son. We'll need you for this. We were worried about you, Alexandrus. The guys are expecting you outside; they want to hear something encouraging from you.

Alexandrus just turned his head to Dulcinea without saying anything, but she just nodded her head, agreeing with what Colonel Demello was saying. He walked outside from the camp.

Alexandrus: Hmm, it's nice to see you all of you guys again. Well, as everybody knows, we're just one step short of finishing this mission. I'm really happy to have come to this point. I know all of you have been through a lot, but this time I'm not begging you, I'm not imploring you, I'm just asking you one more time—let's finish the Zurdos once and for all. It has been a pleasure to hang around with all of you. We don't know if they're still alive, but it's okay because we'll put an end to this. People from Earth, I've got to tell you one more thing: we will prevail in this battle, you know why? Because we're the Earth's warriors and we want to see our children live in peace again. Let's do it for Earth, for all those soldiers that didn't make it to this point, and the most important thing, God bless us all.

Crowd: Yeah!

They were very euphoric; Alexandrus knew how to give those soldiers some hope, especially being too far from home. Then Colonel
Demello stepped out too.

Colonel Demello: It's been a great honor to be on this quest, and
we've come so far that today is the day. You have been my family the
whole time. I've lost a lot friends in this mission and I just want to say
one thing. Let's finish them off.

Crowd: Yes!

Everybody was hugging each other, and they were so motivated
that they were so close to finishing a long journey of horrendous im-
ages.

Dulcinea: Simbra, you stay with Captain J.J.

Simbra: Promise me that you're coming back.

Dulcinea: I promise you.

She kissed Simbra's forehead.

Colonel Demello: Let's do it, let's go, and God bless us all.

Crowd: Let's go.

Everybody was getting out from the place where they were hidden, and they started walking slowly; then they started to run. Alexandrus, Dulcinea, and Amadeus were the first, heading to the City of Fury that was almost destroyed. In the second group was Colonel Demello, Lt. Dago, Atlacatl, and the rest of the group, around nineteen in all. Our heroes were getting closer. The fog was still there, the ashes were blowing in every directions, and they were running so fast when all of sudden Alexandrus made an important decision.

Alexandrus: Wait, stop.

Dulcinea: What? What's happening?

Alexandrus: There's something strange.

Amadeus: Yeah, I can feel it too.

Dulcinea: But what?

Alexandrus: Don't you see, there's nobody here, there's no casualties.

Colonel Demello, Lt. Dago, and Atlacatl were catching up to our heroes.

Colonel Demello: Why are we stopping?

Alexandrus: There's something that I don't like.

Colonel Demello: What do we do now?

Alexandrus: Just look around, you guys, there's nobody; this city was already empty when we launched our attack.

Amadeus: Now what?

Alexandrus: Let's get out of here and soon.

But when Alexandrus was finishing his phrase, the horror began to show up and the ground was shaking.

Colonel Demello: It's an earthquake.

Dulcinea: Let's get back, get back.

But there was no time to get back; the ground was cracking, and our heroes were in the middle of it.

Alexandrus: What's that?

There was a big fence emerging from the ground, surrounding our heroes. They had no escape; they were trapped inside of those walls.

Alexandrus: Great, now we're trapped.

Colonel Demello: We're prisoners.

Dulcinea: There's got to be a way out.

Then they heard a big siren sound, followed by a trembling noise on the ground.

Lt. Dago: Look, over there.

From beneath the ground came the Topos; then on the top of the wall were the Aguilus, and not far from the wall, finally, it was Atonal. He was just watching how his troops were going to kill our heroes. His troops were just waiting for Atonal's order. Meanwhile our troops were ready to fight too, and that was it—the final showdown was going to get started. The Zurdos started the attack first. Our soldiers also were defending themselves. There was no break and worst of all, there was no way of getting out. Alexandrus, Dulcinea, and Amadeus were quick, and they were finishing them off, Atlacatl, Lt. Dago, and Colonel Demello also were doing their part; their swords were covered in blood, but unfortunately our people were outnumbered by far. That wasn't the only problem at all; they were fighting every inch in the battlefield. Atonal was just watching, and our heroes were having a hard time with the Aguilus. The rain started to fall again and they were in a puddle of mud. It was a very emotional moment in how our warriors from Earth were giving all in the battlefield. They didn't care if they were going to win, they just cared about finishing the Zurdos off. When everything seemed that it would end in an unhappy ending, the miracle happened in Europa. From outside of the wall, Simbra and Captain J.J. launched an attack. They had the device to protect against the invisible Aguiluchos, which weren't invisible at all, but it didn't end there. From another part of the wall there was a big blast that opened part of the fence wall. They were the rebels guided by Chafa.

Dulcinea: Look, somebody fired at those Bird Men.

Alexandrus: It must be Captain J.J.

Amadeus: Look, it's Chafa and his people. Now we're even. Let's get them!

Alexandrus and Dulcinea were looking for the Zurdos's leaders. Atonal and Drucila were getting away from the battlefield. They had an emergency exit, which was a small gate. They knew it was getting out of their hands.

Alexandrus: Look over there, they're getting away, I'm going after them. Stay here.

However Dulcinea was after Drucila too. They ran very fast, and the gate was getting close quickly. They barely made it, but they were by themselves; they were inside of the palace underground. Drucila was waiting for them.

Dulcinea: Go, Alexandrus, I'll take care of her. We don't want Atonal escape from this.

Alexandrus just looked at Drucila and he walked away. Both women were facing each other and then they started to fight. Drucila was carrying two swords, which seemed to be an advantage for her. Drucila's swords in any moment weren't stopping, and they wanted to taste Dulcinea's blood so bad.

Drucila: You intruders from the Blue Land are fools.

Dulcinea: We might be fools but not killers.

Drucila: You helped us to get rid of the Dark Knight, and now we have the Valley of silence, just for us. (laughs)

Meanwhile in the other match, after a long race inside of the palace, finally Alexandrus met Atonal. They were going to start a new epic battle. They were studying each other, ready for the final fight of their lives.

Atonal: You think that you can take away what is ours? We're not going to let that happen here. You invaders from the Blue Land can't steal our customs, our way of life. Our ancestors have fought invaders before, and we carry that in our veins. Now it's time for you and your people to die.

Alexandrus: We have come here for justice and we're not going to let you destroy our beautiful planet.

They were fighting on the shore of a lake with their swords. Atonal was a very intimidating warrior; he had a very muscular body and he was really strong. He knew amazing moves with his sword. Both warriors were very skillful on the battlefield. The minutes were passing by and nobody seemed to be losing the fight, but Atonal had some device around his shoulders that was throwing small spears at Alexandrus like some kind of bullets. He had to run fast—very fast. He was getting tired and looking for some kind of shelter, but it was in vain. He was still running away from Atonal's "spears bullets" but then came the moment that Alexandrus was waiting for; Atonal was out of ammo, so Alexandrus was back in action. From martial arts to the unknown "Opico" technique (this one was used to fight to the death in Europa), both knew the skills. The match was so perfect that there was no time to make mistakes. The battle once again took another turn. Atonal was full of surprises. In back of his head he had another lethal weapon, a sharpened pierce that came from nowhere and that thing was so freaky. Atonal, with a nasty kick, found a way to hug Alexandrus, and with this pierce reaching Alexandrus' face, it seemed that Alexandrus was going to be a victim of Atonal. Alexandrus had to get away, and fast. He had no time to wait; the pierce was getting closer and closer.

On the other side of the battle, Amadeus, Colonel Demello, and company had won their matches, but they couldn't get in. The gate was already closed. Dulcinea was the last one to cross it, so it was up to Alexandrus and Dulcinea.

Colonel Demello: It's closed.

Amadeus: I know.

Atlacatl: I hope they can get away from this.

Amadeus: I'm sure they will.

Dulcinea and Drucila were having their own combat. Drucila was trying to put Dulcinea in a container which the Zurdos used to melt the Super Eurotanius. She was in between.

Drucila: Now, it's time for you to die.

Dulcinea: But not here.

With a magnificent move, Dulcinea avoided Drucila's swords and she had Drucila in her place.

Dulcinea: The time to die is for you.

She kicked Drucila into the container, and she was screaming. There was no way to get out, so it was the only way to defeat the "Metal-woman." She melted.

Dulcinea: Now you can take your two faces somewhere else.

She left; she was going to help Alexandrus.

Alexandrus had to get rid of that pierce; he was bleeding in one of his shoulders, and there was no time, he had to get away. He had to step up, he had to get over, and he had to show why he was one of the chosen from Earth.

With his last breath of life, he pulled out with his legs (he had strong legs from playing soccer). Then he grabbed his sword and...

Alexandrus: No more jokes, Atonal, your horrible ruling days are over.

He thrust his sword in Atonal's neck, once and for all, Alexandrus had defeated the leader of the Zurdos, but not before Alexandrus went down too. At that moment Dulcinea showed up. After seeing Alexandrus down, she went to him right away.

Dulcinea: Are you all right? Answer me, please.

There was no answer. She started to give artificial respiration, pushing on Alexandrus's chest.

Dulcinea: Come on, you're not going to die here, please God!

She started to cry and then the miracle happened once again. Alexandrus reacted little by little.

Alexandrus: I'm okay. Our mission is finished here. Let's go home.

He was tired.

Dulcinea: (screaming to the sky) Thank you, God!

They were walking toward the entrance of the city, and when they opened the gate, they found a big surprise.

Crowd: Yeah.

Alexandrus: Now you can live in peace, people of Europa, no more terror in this place. Atonal is gone, you have a bright future, and our mission is done here. No more days of massacres, no more famine, no more destruction in this place. The Zurdos are gone. I want to ask for a minute of silence for all those who didn't make it. Just look around you—now we have a new chapter in our life. Thank you for your help. Without it, we wouldn't have done it. Now we're going home, and remember, there will come better days.

Crowd: Yeah.

Everybody was hugging each other, singing, and dancing. Europa finally had its freedom and our heroes were thinking of how to get home, but that is a different story. And remember, there will come better days.